Basic Training in Chemistry

Basic Training in Chemistry

Steven L. Hoenig
Ridgewood, New York

Kluwer Academic / Plenum Publishers
New York ● Boston ● Dordrecht ● London ● Moscow

Library of Congress Cataloging-in-Publication Data

Hoenig, Steven L., 1957–
    Basic training in chemistry/Steven L. Hoenig.
        p. cm.
    Includes bibliographical references and index.
    ISBN 0-306-46546-9
        1. Chemistry. I. Title.

    QD31.2 .H58 2001
    540—dc21

                                            00-067108

ISBN 0-306-46546-9

©2001 Kluwer Academic / Plenum Publishers, New York
233 Spring Street, New York, New York 10013

http://www.wkap.nl/

10 9 8 7 6 5 4 3 2 1

A C.I.P. record for this book is available from the Library of Congress

Printed in the United States of America

To Lena and Alan

Preface

This book was written as a quick reference to the many different concepts and ideas encountered in chemistry. Most books these days go into a detailed explanation of one subject and go no further. This is simply an attempt to present briefly some of the various subjects that make up the whole of chemistry. The different subjects covered include general chemistry, inorganic chemistry, organic chemistry, and spectral analysis. The material is brief, but hopefully detailed enough to be of use. Keep in mind that the material is written for a reader who is familiar with the subject of chemistry. It has been the author's intention to present in one ready source several disciplines that are used and referred to often.

This book was written not to be a chemistry text unto itself, but rather as a supplement that can be used repeatedly throughout a course of study and thereafter. This does not preclude it from being used by others that would find it useful as a reference source as well.

Having kept this in mind during its preparation, the material is presented in a manner in which the reader should have some knowledge of the material. Only the basics are stated because a detailed explanation was not the goal but rather to present a number of chemical concepts in one source.

The first chapter deals with material that is commonly covered in almost every first year general chemistry course. The concepts are presented in, I hope, a clear and concise manner. No detailed explanation of the origin of the material or problems are presented. Only that which is needed to understand the concept is stated. If more detailed explanation is needed any general chemistry text would suffice. And if examples are of use, any review book could be used.

The second chapter covers inorganic chemistry. Those most commonly encountered concepts are presented, such as, coordination numbers, crystal systems, and ionic crystals. More detailed explanation of the coordination encountered in bonding of inorganic compounds requires a deeper explanation then this book was intended for.

Chapter three consists mostly of organic reactions listed according to their preparation and reactions. The mechanisms of the various reactions are not discussed since there are numerous texts which are devoted to the subject. A section is devoted to the concept of isomers since any treatment of organic chemistry must include an understanding it. A section on polymer structures is also presented since it is impossible these days not to come across some discussion of it.

The fourth chapter covers instrumental analysis. No attempt is made to explain the inner workings of the different instruments or the mechanisms by which various spectra is produced. The material listed is for the use by those that are familiar with the different type of spectra encountered in the instrumental analysis of chemical compounds. The tables and charts would be useful for the interpretation of various spectra generated in the course of analyzing a chemical substance. Listed are tables that would be useful as for the interpretation of ultra-violet (uv), infra-red (ir), nuclear magnetic resonance (nmr) and mass spectroscopy (ms) spectra.

Chapter five consists of physical constants and unit measurements that are commonly encountered throughout the application of chemistry.

Chapter six contains certain mathematical concepts that are useful to have when reviewing or working with certain concepts encountered in chemistry.

Steven L. Hoenig

Acknowledgements

I wish to express my deepest gratitude to Richard Leff, without whose faith and belief in me this book would not have been possible. And I wish to express my thanks to the following Richard Kolodkin, Robert Glatt, Cindy and Harriet Cuccias, and Anthony Woll, who know what the meaning of friendship is.

Contents

Chapter 1

General Chemistry

1.1   Atomic Structure and the Periodic Table

1.1.1   Constituents of the Atom

The atom of any element consists of three basic types of particles...the electron (a negatively charged particle), the proton (a positively charged particle), and the neutron (a neutrally charged particle). The protons and neutrons occupy the nucleus while the electrons are outside of the nucleus. The protons and neutrons contribute very little to the total volume but account for the majority of the atom's mass. However, the atoms volume is determined the electrons, which contribute very little to the mass. Table 1.1 summarizes the properties of these three particles.

Table 1.1. Properties of the Proton, Electron, and Neutron

Particle	Mass	Electric Charge	Unit Charge
Proton	1.672×10^{-24} g	$+1.602 \times 10^{-19}$ coulomb	$+1$
Electron	9.108×10^{-28} g	-1.602×10^{-19} coulomb	-1
Neutron	1.675×10^{-24} g	0	0

The **atomic number** (Z) of an element is the number of protons within the nucleus of an atom of that element. In a neutral atom, the number of protons and electrons are equal and the atomic number also indicates the number of electrons.

The **mass number** (A) is the sum of the protons and neutrons present in the atom. The number of neutrons can be determined by (A - Z). The symbol for denoting the atomic number and mass number for an element X is as follows:

$$^A_Z X$$

Atoms that have the same atomic number (equal number of protons) but different atomic masses (unequal number of neutrons) are referred to as isotopes. For example, carbon consists of two isotopes, carbon-12 and carbon-13:

$$^{12}_6 C \quad ^{13}_6 C$$

The **atomic mass unit** (amu) is defined as 1/12 the mass of a carbon-12 isotope. The relative **atomic mass** of an element is the weighted average of the isotopes relative to 1/12 of the carbon-12 isotope. For example, the atomic mass of neon is 20.17 amu and is calculated from the following data: neon-19 (amu of 19.99245, natural abundance of 90.92%), neon-20 (amu of 20.99396, natural abundance of 0.260%) and neon-21 (amu of 21.99139, natural abundance of 8.82%):

a.m. neon=(19.99245*0.9092)+(20.99396*0.00260)+(21.99139*0.0882)
= 20.17 amu

The relative **molecular mass** is the sum of the atomic masses for each atom in the molecule. For $H_2SO_4 = (1 * 2) + 32 + (16 * 4) = 98$.

The **mole** (mol) is simply a unit of quantity, it represents a certain amount of material, i.e. atoms or molecules. The numerical value of one mole is 6.023×10^{23} and is referred to as **Avogadro's number**. The mole is defined as the mass, in grams, equal to the atomic mass of an element or molecule. Therefore, 1 mole of carbon weighs 12 grams and contains 6.023×10^{23} carbon atoms. The following formula can be used to find the number of moles:

$$moles = \frac{mass \ in \ grams}{atomic \ (or \ molecular) \ mass}$$

1.1.2 Quantum Numbers

From quantum mechanics a set of equations called wave equations are obtained. A series of solutions to these equations, called wave functions, gives the four quantum numbers required to describe the placement of the electrons in the hydrogen atom or in other atoms.

The **principal quantum number,** n, determines the energy of an orbital and has a value of n = 1, 2, 3, 4, ...

The **angular momentum quantum number**, λ, determines the "shape" of the orbital and has a value of 0 to (n - 1) for every value of n.

The **magnetic quantum number**, ml, determines the orientation of the orbital in space and has a value of $-\lambda$ to $+\lambda$.

The **electron spin quantum number**, m$_S$, determines the magnetic field generated by the electron and has a value of $-\frac{1}{2}$ or $+\frac{1}{2}$.

Table 1.2. Quantum Numbers and Electron Distribution

Shell	Principal quantum number n	Angular momentum quantum number λ	Orbital designation*	Magnetic quantum number m$_\lambda$	Spin quantum number m$_S$	Total number of electrons per orbital
K	1	0	s	0	$-\frac{1}{2}, +\frac{1}{2}$	2
L	2	0	s	0	$-\frac{1}{2}, +\frac{1}{2}$	2
		1	p$_x$	-1	$-\frac{1}{2}, +\frac{1}{2}$	
			p$_y$	0	$-\frac{1}{2}, +\frac{1}{2}$	6
			p$_z$	+1	$-\frac{1}{2}, +\frac{1}{2}$	
M	3	0	s	0	$-\frac{1}{2}, +\frac{1}{2}$	2
		1	p$_x$	-1	$-\frac{1}{2}, +\frac{1}{2}$	
			p$_y$	0	$-\frac{1}{2}, +\frac{1}{2}$	6
			p$_z$	+1	$-\frac{1}{2}, +\frac{1}{2}$	
		2	d$_{xy}$	-2	$-\frac{1}{2}, +\frac{1}{2}$	
			d$_{xz}$	-1	$-\frac{1}{2}, +\frac{1}{2}$	
			d$_{yz}$	0	$-\frac{1}{2}, +\frac{1}{2}$	10
			d$_{z2}$	+1	$-\frac{1}{2}, +\frac{1}{2}$	
			d$_{x2-y2}$	+2	$-\frac{1}{2}, +\frac{1}{2}$	

The following is a summary in which the quantum numbers are used to fill the atomic orbitals:

1. No two electrons can have the same four quantum numbers. This is the Pauli exclusion principle.
2. Orbitals are filled in the order of increasing energy.
3. Each orbital can only be occupied by a maximum of two electrons and must have different spin quantum numbers (opposite spins).
4. The most stable arrangement of electrons in orbitals is the one that has the greatest number of equal spin quantum numbers (parallel spins). This is Hund's rule.

Also note that the energy of an electron also depends on the angular momentum quantum number as well as the principal quantum number. Therefore the order that the orbitals get filled does not strictly follow the principal quantum number. The order in which orbitals are filled is given in figure 1.1.

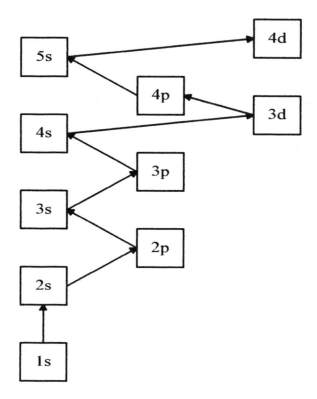

Figure 1.1 Filling Order and Relative Energy Levels of Orbitals

1.1.3   Atomic Orbitals

The quantum numbers mentioned earlier were obtained as solutions to a set of wave equations. These wave equations cannot tell precisely where an electron is at any given moment or how fast it is moving. But rather it states the probability of finding the electron at a particular place. An orbital is a region of space where the electron is most likely to be found. An orbital has no definite boundary to it, but can be thought of as a cloud with a specific shape. Also, the orbital is not uniform throughout, but rather densest in the region where the probability of finding the electron is highest.

The shape of an orbital represents 90% of the probability of finding the electron within that space. As the quantum numbers change so do the shapes and direction of the orbitals. Figure 1.2 show the shapes for principal quantum number n = 1, 2, and 3.

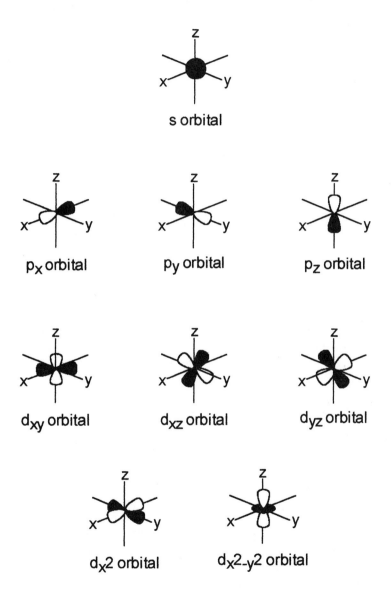

Figure 1.2 Representation of Atomic Orbitals

Another type of orbital is one that originates from the mixing of the different atomic orbitals and is called a hybrid orbital. Hybridization (mixing) of atomic orbitals results in a new set of orbitals with different shapes and orientations. The orbitals are designated according to which of the separate atomic orbitals have been mixed. For instance, an s orbital mixing with a single p orbital is designated sp. An s orbital mixing with two separate p orbitals ($p_x + p_y$ or $p_x + p_z$ or $p_y + p_z$) is designated sp^2 and a s orbital mixing with three separate p orbitals is designated sp^3. Combinations of other orbitals can occur as well. Table 1.3 lists some of the possible hybrid orbitals.

Table 1.3. Hybrid Orbitals

Hybrid orbital	Atomic orbitals hybridized	Number of hybrid orbitals	Angle of hybrid orbital	Geometry	Example
sp dp	s + p d + p	2 2	180° 180°	————	CO_2 $HgCl_2$
sp^2	s + $p_x + p_y$	3	120°		PCl_3 SO_3 H_2CO
sp^3	s + $p_x + p_y + p_z$	4	109°28'		CH_4 $AlCl_4$
dsp² d²p²	1+ px + py + d_{x2-y2} $d_{x2-y2} + d_z2 +$ $p_x + p_y$	4 4	90° 90°		XeF_4
dsp^3	s + $p_x + p_y + p_z +$ d_{x2-y2}	5	90° 120° 180°		PCl_5 PF_5
d^2sp^3	s + $p_x + p_y + p_z +$ $d_{x2-y2} + d_z2$	6	90° 180°		SF_6 XeF_6 MoF_6

1.1.4 Electronic Configuration of the Elements

Table 1.4. Electronic Configuration of the Elements

Shells		K	L		M			N				O				P				Q
Sub-Levels		1s	2s	2p	3s	3p	3d	4s	4p	4d	4f	5s	5p	5d	5f	6s	6p	6d	6f	7s
1	Hydrogen	1																		
2	Helium	2																		
3	Lithium	2	1																	
4	Beryllium	2	2																	
5	Boron	2	2	1																
6	Carbon	2	2	2																
7	Nitrogen	2	2	3																
8	Oxygen	2	2	4																
9	Fluorine	2	2	5																
10	Neon	2	2	6																
11	Sodium	2	2	6	1															
12	Magnesium	2	2	6	2															
13	Aluminum	2	2	6	2	1														
14	Silicon	2	2	6	2	2														
15	Phosphorus	2	2	6	2	3														
16	Sulfur	2	2	6	2	4														
17	Chlorine	2	2	6	2	5														
18	Argon	2	2	6	2	6														
19	Potassium	2	2	6	2	6		1												
20	Calcium	2	2	6	2	6		2												
21	Scandium	2	2	6	2	6	1	2												
22	Titanium	2	2	6	2	6	2	2												
23	Vanadium	2	2	6	2	6	3	2												
24	Chromium	2	2	6	2	6	5	1												
25	Manganese	2	2	6	2	6	5	2												
26	Iron	2	2	6	2	6	6	2												
27	Cobalt	2	2	6	2	6	7	2												
28	Nickel	2	2	6	2	6	8	2												
29	Copper	2	2	6	2	6	10	1												
30	Zinc	2	2	6	2	6	10	2												
31	Gallium	2	2	6	2	6	10	2	1											
32	Germanium	2	2	6	2	6	10	2	2											
33	Arsenic	2	2	6	2	6	10	2	3											
34	Selenium	2	2	6	2	6	10	2	4											
35	Bromine	2	2	6	2	6	10	2	5											
36	Krypton	2	2	6	2	6	10	2	6											
37	Rubidium	2	2	6	2	6	10	2	6			1								
38	Strontium	2	2	6	2	6	10	2	6			2								
39	Yttrium	2	2	6	2	6	10	2	6	1		2								
40	Zirconium	2	2	6	2	6	10	2	6	2		2								
41	Niobium	2	2	6	2	6	10	2	6	4		1								
42	Molybdenum	2	2	6	2	6	10	2	6	5		1								

Table 1.4. (Continued).

	Shells	K	L		M			N				O				P				Q
	Sub-Levels	1s	2s	2p	3s	3p	3d	4s	4p	4d	4f	5s	5p	5d	5f	6s	6p	6d	6f	7s
43	Technetium	2	2	6	2	6	10	2	6	6		1								
44	Ruthenium	2	2	6	2	6	10	2	6	7		1								
45	Rhodium	2	2	6	2	6	10	2	6	8		1								
46	Palladium	2	2	6	2	6	10	2	6	10										
47	Silver	2	2	6	2	6	10	2	6	10		1								
48	Cadmium	2	2	6	2	6	10	2	6	10		2								
49	Indium	2	2	6	2	6	10	2	6	10		1	2							
50	Tin	2	2	6	2	6	10	2	6	10		2	2							
51	Antimony	2	2	6	2	6	10	2	6	10		2	3							
52	Tellurium	2	2	6	2	6	10	2	6	10		2	4							
53	Iodine	2	2	6	2	6	10	2	6	10		2	5							
54	Xenon	2	2	6	2	6	10	2	6	10		2	6							
55	Cesium	2	2	6	2	6	10	2	6	10		2	6			1				
56	Barium	2	2	6	2	6	10	2	6	10		2	6			2				
57	Lanthanum	2	2	6	2	6	10	2	6	10		2	6	1		2				
58	Cerium	2	2	6	2	6	10	2	6	10	2	2	6			2				
59	Praseodymium	2	2	6	2	6	10	2	6	10	3	2	6			2				
60	Neodymium	2	2	6	2	6	10	2	6	10	4	2	6			2				
61	Promethium	2	2	6	2	6	10	2	6	10	5	2	6			2				
62	Samarium	2	2	6	2	6	10	2	6	10	6	2	6			2				
63	Europium	2	2	6	2	6	10	2	6	10	7	2	6			2				
64	Gadolinium	2	2	6	2	6	10	2	6	10	7	2	6	1		2				
65	Terbium	2	2	6	2	6	10	2	6	10	9	2	6			2				
66	Dysprosium	2	2	6	2	6	10	2	6	10	10	2	6			2				
67	Holium	2	2	6	2	6	10	2	6	10	11	2	6			2				
68	Erbium	2	2	6	2	6	10	2	6	10	12	2	6			2				
69	Thulium	2	2	6	2	6	10	2	6	10	13	2	6			2				
70	Ytterbium	2	2	6	2	6	10	2	6	10	14	2	6			2				
71	Lutetium	2	2	6	2	6	10	2	6	10	14	2	6	1		2				
72	Hafnium	2	2	6	2	6	10	2	6	10	14	2	6	2		2				
73	Tantalium	2	2	6	2	6	10	2	6	10	14	2	6	3		2				
74	Tungsten	2	2	6	2	6	10	2	6	10	14	2	6	4		2				
75	Rhenium	2	2	6	2	6	10	2	6	10	14	2	6	5		2				
76	Osmium	2	2	6	2	6	10	2	6	10	14	2	6	6		2				
77	Iridium	2	2	6	2	6	10	2	6	10	14	2	6	9						
78	Platinum	2	2	6	2	6	10	2	6	10	14	2	6	9		1				
79	Gold	2	2	6	2	6	10	2	6	10	14	2	6	10		1				
80	Mercury	2	2	6	2	6	10	2	6	10	14	2	6	10		2				
81	Thallium	2	2	6	2	6	10	2	6	10	14	2	6	10		2	1			
82	Lead	2	2	6	2	6	10	2	6	10	14	2	6	10		2	2			
83	Bismuth	2	2	6	2	6	10	2	6	10	14	2	6	10		2	3			
84	Polonium	2	2	6	2	6	10	2	6	10	14	2	6	10		2	4			
85	Astatine	2	2	6	2	6	10	2	6	10	14	2	6	10		2	5			
86	Radon	2	2	6	2	6	10	2	6	10	14	2	6	10		2	6			

Table 1.4. (Continued)

	Shells	K	L		M			N				O				P				Q
	Sub-Levels	1s	2s	2p	3s	3p	3d	4s	4p	4d	4f	5s	5p	5d	5f	6s	6p	6d	6f	7s
87	Francium	2	2	6	2	6	10	2	6	10	14	2	6	10		2	6			1
88	Radium	2	2	6	2	6	10	2	6	10	14	2	6	10		2	6			2
89	Actinium	2	2	6	2	6	10	2	6	10	14	2	6	10		2	6	1		2
90	Thorium	2	2	6	2	6	10	2	6	10	14	2	6	10		2	6	2		2
91	Protactinium	2	2	6	2	6	10	2	6	10	14	2	6	10	2	2	6	1		2
92	Uranium	2	2	6	2	6	10	2	6	10	14	2	6	10	3	2	6	1		2
93	Neptunium	2	2	6	2	6	10	2	6	10	14	2	6	10	4	2	6	1		2
94	Plutonium	2	2	6	2	6	10	2	6	10	14	2	6	10	6	2	6			2
95	Americium	2	2	6	2	6	10	2	6	10	14	2	6	10	7	2	6			2
96	Curium	2	2	6	2	6	10	2	6	10	14	2	6	10	7	2	6	1		2
97	Berkelium	2	2	6	2	6	10	2	6	10	14	2	6	10	9	2	6			2
98	Californium	2	2	6	2	6	10	2	6	10	14	2	6	10	10	2	6			2
99	Einsteinium	2	2	6	2	6	10	2	6	10	14	2	6	10	11	2	6	1		2
100	Fermium	2	2	6	2	6	10	2	6	10	14	2	6	10	12	2	6	1		2
101	Mendelevium	2	2	6	2	6	10	2	6	10	14	2	6	10	13	2	6	1		2
102	Nobelium	2	2	6	2	6	10	2	6	10	14	2	6	10	14	2	6	1		2
103	Lawrencium	2	2	6	2	6	10	2	6	10	14	2	6	10	14	2	6	1		2
104	Unnilquadium	2	2	6	2	6	10	2	6	10	14	2	6	10	14	2	6	2		2
105	Unnilpentium	2	2	6	2	6	10	2..6		10	14	2	6	10	14	2	6	3		2
106	Unnilhexium	2	2	6	2	6	10	2	6	10	14	2	6	10	14	2	6	4		2
107	Unnilseptium	2	2	6	2	6	10	2	6	10	14	2	6	10	14	2	6	5		2
108																				

1.1.5   Periodic Table of Elements

Table 1.5. Periodic Table of Elements

IA	IIA	IIIB	IVB	VB	VIB	VIIB	VIIIB	VIIIB	VIIIB	IB	IIB	IIIA	IVA	VA	VIA	VIIA	VIIIA
1 **H** 1.008																	2 **He** 4.0026
3 **Li** 6.939	4 **Be** 9.012											5 **B** 10.811	6 **C** 12.011	7 **N** 14.0067	8 **O** 15.9994	9 **F** 18.998	10 **Ne** 20.179
11 **Na** 22.991	12 **Mg** 24.312											13 **Al** 26.982	14 **Si** 28.086	15 **P** 30.974	16 **S** 78.96	17 **Cl** 35.453	18 **Ar** 39.948
19 **K** 39.102	20 **Ca** 40.08	21 **Sc** 44.956	22 **Ti** 47.90	23 **V** 50.942	24 **Cr** 51.996	25 **Mn** 54.938	26 **Fe** 55.847	27 **Co** 58.933	28 **Ni** 58.71	29 **Cu** 63.546	30 **Zn** 65.37	31 **Ga** 69.72	32 **Ge** 72.59	33 **As** 74.922	34 **Se** 78.96	35 **Br** 79.909	36 **Kr** 83.80
37 **Rb** 85.47	38 **Sr** 87.62	39 **Y** 88.905	40 **Zr** 91.22	41 **Nb** 92.906	42 **Mo** 96.94	43 **Tc** 98.906	44 **Ru** 101.07	45 **Rh** 102.905	46 **Pd** 106.4	47 **Ag** 107.868	48 **Cd** 112.40	49 **In** 114.82	50 **Sn** 118.69	51 **Sb** 121.75	52 **Te** 127.60	53 **I** 126.904	54 **Xe** 131.30
55 **Cs** 132.905	56 **Ba** 137.34	57 **La** 138.91	72 **Hf** 178.49	73 **Ta** 180.948	74 **W** 183.85	75 **Re** 186.2	76 **Os** 190.2	77 **Ir** 192.2	78 **Pt** 195.09	79 **Au** 196.967	80 **Hg** 200.59	81 **Tl** 204.37	82 **Pb** 207.19	83 **Bi** 208.980	84 **Po** (209)	85 **At** (210)	86 **Rn** (222)
87 **Fr** (223)	88 **Ra** 226.02	89 **Ac** (227)	104 **Unq** (261)	105 **Unp** (262)	106 **Unh** (263)	107 **Uns** (261)											

58 **Ce** 140.12	59 **Pr** 140.90	60 **Nd** 144.24	61 **Pm** (145)	62 **Sm** 150.35	63 **Eu** 151.96	64 **Gd** 157.25	65 **Tb** 158.924	66 **Dy** 162.50	67 **Ho** 164.930	68 **Er** 167.26	69 **Tm** 168.934	70 **Yb** 173.04	71 **Lu** 174.97
90 **Th** 232.038	91 **Pa** 231.03	92 **U** 238.03	93 **Np** 237.04	94 **Pu** (244)	95 **Am** (243)	96 **Cm** (247)	97 **Bk** (247)	98 **Cf** (251)	99 **Es** (254)	100 **Fm** (257)	101 **Md** (258)	102 **No** (259)	103 **Lr** (260)

1.1.6   Ionization Energy

Ionization energy is the minimum amount of energy needed to remove an electron from a gasous atom or ion, and is expressed in electron volts (eV). Going across the periodic table the I.E. increases due to the fact that the principal energy level (principal quantum number) remains the same while the number of electrons increase, thereby enhancing the electrostatic attraction between the protons in the nulceus and the electrons. Going down the table the I.E. decreases because the outer electrons are now further from the nucleus and the protons.

Table 1.6 shows the first ionization energy for the elements.

$$M \text{ (gas)} \rightarrow M^+ \text{ (gas)} + e^-$$

Table 1.6. First Ionization Energy (in eV)

Z	Element	I.E.	Z	Element	I.E.	Z	Element	I.E.
1	Hydrogen	13.59	30	Zinc	9.39	59	Praseodymium	5.40
2	Helium	24.58	31	Gallium	6.00	60	Neodymium	5.49
3	Lithium	5.39	32	Germanium	7.88	61	Promethium	5.55
4	Beryllium	9.32	33	Arsenic	9.81	62	Samarium	5.61
5	Boron	8.30	34	Selenium	9.75	63	Europium	5.64
6	Carbon	11.26	35	Bromine	11.84	64	Gadolinium	6.26
7	Nitrogen	14.53	36	Krypton	14.00	65	Terbium	5.89
8	Oxygen	13.61	37	Rubidium	4.18	66	Dysprosium	5.82
9	Fluorine	17.42	38	Strontium	5.69	67	Holium	5.89
10	Neon	21.56	39	Yttrium	6.38	68	Erbium	5.95
11	Sodium	5.14	40	Zirconium	6.84	69	Thulium	6.03
12	Magnesium	7.64	41	Niobium	6.88	70	Ytterbium	6.04
13	Aluminum	5.98	42	Molybdenum	7.10	71	Lutetium	5.32
14	Silicon	8.15	43	Technetium	7.28	72	Hafnium	7.00
15	Phosphorus	10.48	44	Ruthenium	7.36	73	Tantalium	7.88
16	Sulfur	10.36	45	Rhodium	7.46	74	Tungsten	7.98
17	Chlorine	12.97	46	Palladium	8.33	75	Rhenium	7.87
18	Argon	15.76	47	Silver	7.57	76	Osmium	8.73
19	Potassium	4.34	48	Cadmium	8.99	77	Iridium	9.1
20	Calcium	6.11	49	Indium	5.79	78	Platinum	8.96
21	Scandium	6.54	50	Tin	7.34	79	Gold	9.22
22	Titanium	6.82	51	Antimony	8.64	80	Mercury	10.43
23	Vanadium	6.74	52	Tellurium	9.01	81	Thallium	6.11
24	Chromium	6.87	53	Iodine	10.45	82	Lead	7.42
25	Manganese	7.43	54	Xenon	12.13	83	Bismuth	7.29
26	Iron	7.87	55	Cesium	3.89	84	Polonium	8.43
27	Cobalt	7.86	56	Barium	5.21	85	Astatine	9.5
28	Nickel	7.63	57	Lanthanum	5.61	86	Radon	10.75
29	Copper	7.72	58	Cerium	6.54			

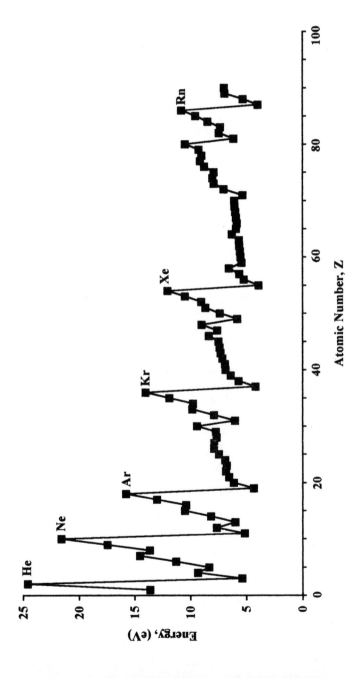

Figure 1.3. First ionization energy

1.1.7   Electronegativity

Electronegativity (X) is the relative attraction of an atom for an electron in a covalent bond. But due to the complexity of a covalent bond it is not possible to define precise electronegativity values. Originally the element fluorine, whose atoms have the greatest attraction for an electron, was given an arbitrary value of 4.0. All other electronegativity values are based on this.

Note that the greater the difference in electronegativities the more ionic in nature is the bond, and the smaller the difference the more covalent is the bond.

Going across the periodic table the electronegativity increases because the principal energy level remain the same and the electrostatic attraction increases. The atoms also have a desire to have the most stable configuration which is that of the noble gas configuration. Going down the table the electronegativity decreases due to the increased distance from the nucleus.

Table 1.7 lists relative electronegativities for the elements.

Table 1.7. Relative Electronegativities

Z	Element	X	Z	Element	X	Z	Element	X
1	Hydrogen	2.2	27	Cobalt	1.9	53	Iodine	2.5
2	Helium	---	28	Nickel	1.9	54	Xenon	---
3	Lithium	1.0	29	Copper	1.9	55	Cesium	0.7
4	Beryllium	1.5	30	Zinc	1.6	56	Barium	0.9
5	Boron	2.0	31	Gallium	1.6			
6	Carbon	2.5	32	Germanium	1.8	72	Hafnium	1.3
7	Nitrogen	3.0	33	Arsenic	2.0	73	Tantalium	1.5
8	Oxygen	3.5	34	Selenium	2.4	74	Tungsten	1.7
9	Fluorine	4.0	35	Bromine	2.8	75	Rhenium	1.9
10	Neon	---	36	Krypton	---	76	Osmium	2.2
11	Sodium	0.9	37	Rubidium	0.8	77	Iridium	2.2
12	Magnesium	1.2	38	Strontium	1.0	78	Platinum	2.2
13	Aluminum	1.5	39	Yttrium	1.2	79	Gold	2.4
14	Silicon	2.8	40	Zirconium	1.4	80	Mercury	1.9
15	Phosphorus	2.1	41	Niobium	1.6	81	Thallium	1.8
16	Sulfur	2.5	42	Molybdenum	1.8	82	Lead	1.9
17	Chlorine	3.0	43	Technetium	1.9	83	Bismuth	1.9
18	Argon	---	44	Ruthenium	2.2	84	Polonium	2.0
19	Potassium	0.8	45	Rhodium	2.2	85	Astatine	2.2
20	Calcium	1.0	46	Palladium	2.2	86	Radon	---
21	Scandium	1.3	47	Silver	1.9	87	Francium	0.7
22	Titanium	1.5	48	Cadmium	1.7	88	Radium	0.9
23	Vanadium	1.6	49	Indium	1.7	89	Actinium	1.1
24	Chromium	1.6	50	Tin	1.8	90	Thorium	1.3
25	Manganese	1.5	51	Antimony	1.9	91	Protactinium	1.4
26	Iron	1.8	52	Tellurium	2.1	92	Uranium	1.4

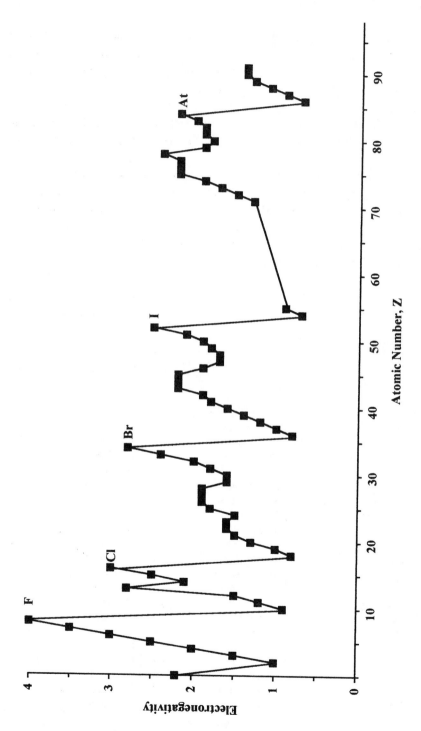

Figure 1.4. Electronegativity

1.1.8   Radius of Atoms

The radius of an atom can be estimated by taking half the distance between the nucleus of two of the same atoms. For example, the distance between the nuclei of I_2 is 2.66 Å, half that distance would be the radius of atomic iodine or 1.33 Å. Using this method the atomic radius of nearly all the elements can be estimated.

Note that going across the periodic table, the atomic radius decreases. This is due to the fact that the principal energy level (principal quantum number) remains the same, but the number of electrons increase. The increase in the number of electrons causes an increase in the electrostatic attraction which causes the radius to decrease. However, going down the periodic table the principal energy level increases and hence the atomic radius increases.

Table 1.8 lists the atomic radii of some of the elements.

Table 1.8. Atomic Radii (in Å)

Z	Element	X	Z	Element	X	Z	Element	X
1	Hydrogen	0.37	25	Manganese	1.29	49	Indium	1.62
2	Helium	----	26	Iron	1.26	50	Tin	1.40
3	Lithium	1.52	27	Cobalt	1.26	51	Antimony	1.41
4	Beryllium	1.12	28	Nickel	1.24	52	Tellurium	1.37
5	Boron	0.88	29	Copper	1.28	53	Iodine	1.33
6	Carbon	0.77	30	Zinc	1.33	54	Xenon	----
7	Nitrogen	0.70	31	Gallium	1.22	55	Cesium	2.62
8	Oxygen	0.66	32	Germanium	1.22	56	Barium	2.17
9	Fluorine	0.64	33	Arsenic	1.21			
10	Neon	----	34	Selenium	1.17	72	Hafnium	1.57
11	Sodium	1.86	35	Bromine	1.14	73	Tantalium	1.43
12	Magnesium	1.60	36	Krypton	----	74	Tungsten	1.37
13	Aluminum	1.43	37	Rubidium	2.41	75	Rhenium	1.37
14	Silicon	1.17	38	Strontium	2.15	76	Osmium	1.34
15	Phosphorus	1.10	39	Yttrium	1.80	77	Iridium	1.35
16	Sulfur	1.04	40	Zirconium	1.57	78	Platinum	1.38
17	Chlorine	0.99	41	Niobium	1.43	79	Gold	1.44
18	Argon	----	42	Molybdenum	1.36	80	Mercury	1.50
19	Potassium	2.31	43	Technetium	1.30	81	Thallium	1.71
20	Calcium	1.97	44	Ruthenium	1.33	82	Lead	1.75
21	Scandium	1.60	45	Rhodium	1.34	83	Bismuth	1.46
22	Titanium	1.46	46	Palladium	1.38	84	Polonium	1.40
23	Vanadium	1.31	47	Silver	1.44	85	Astatine	1.40
24	Chromium	1.25	48	Cadmium	1.49	86	Radon	----

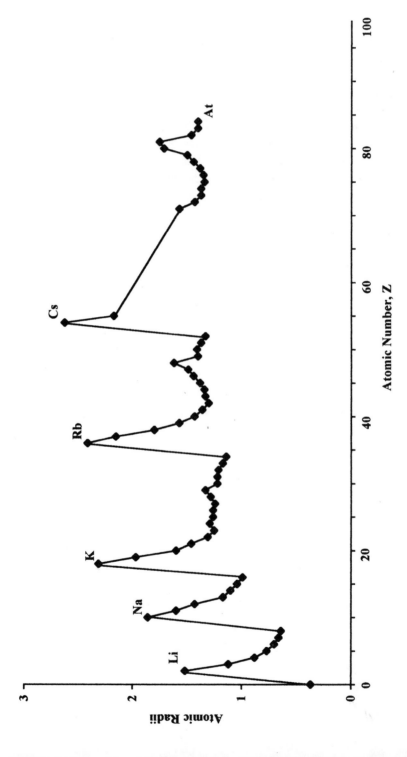

Figure 1.5. Atomic Radii

1.1.9 Atomic Weights

Table 1.9. Atomic Weights

Name	Symbol	Atomic Wt	Name	Symbol	Atomic Wt
Actinium	Ac	227.0278	Neodymium	Nd	144.24
Aluminum	Al	26.9815	Neon	Ne	20.183
Americium	Am	(243.0614)	Neptunium	Np	237.0482
Antimony	Sb	121.75	Nickel	Ni	58.71
Argon	Ar	39.948	Niobium	Nb	92.906
Arsenic	As	74.9216	Nitrogen	N	14.0067
Astatine	At	(209.9871)	Nobelium	No	(259.1009)
Barium	Ba	137.34	Osmium	Os	190.2
Berkelium	Bk	(247.0703)	Oxygen	O	15.9994
Beryllium	Be	9.0122	Palladium	Pd	106.4
Bismuth	Bi	208.980	Phosphorus	P	30.9738
Boron	B	10.811	Platinum	Pt	195.09
Bromine	B	79.909	Plutonium	Pu	(244.0642)
Cadium	Cd	12.401	Polonium	Po	(208.9824)
Calcium	Ca	40.08	Potassium	K	39.102
Californium	Cf	(251.0796)	Praseodymium	Pr	140.907
Carbon	C	12.01115	Promethium	Pm	(144.9127)
Cerium	Ce	140.12	Protactinium	Pa	231.0359
Cesium	Cs	132.905	Radium	Ra	226.0254
Chlorine	Cl	35.453	Radon	Rn	(222.0176)
Chromium	Cr	51.996	Rhenium	Re	186.2
Cobalt	Co	58.9332	Rhodium	Rh	102.905
Copper	Cu	3.546	Rubidium	Rb	85.47
Curium	Cm	(247.0703)	Ruthenium	Ru	101.07
Dysprosium	Dy	162.50	Samarium	Sm	150.35
Einsteinium	Es	(252.083)	Scandium	Sc	44.956
Erbium	Er	167.26	Selenium	Se	78.96
Europium	Eu	151.96	Silicon	Si	28.086
Fermium	Fm	(257.0951)	Silver	Ag	107.870
Fluorine	F	18.9984	Sodium	Na	22.9898
Francium	Fr	(223.0197)	Strontium	Sr	87.62
Gadolinium	Gd	157.25	Sulfur	S	32.064
Gallium	Ga	69.72	Tantalum	Ta	180.948
Germanium	Ge	72.59	Technetium	Tc	98.906
Gold	Au	196.967	Tellurium	Te	127.60
Hafnium	Hf	178.49	Terbium	Tb	158.924
Helium	He	4.0026	Thallium	Tl	204.37
Holmium	Ho	164.930	Thorium	Th	232.038
Hydrogen	H	1.00797	Thulium	Tm	168.934
Indium	In	114.82	Tin	Sn	118.69
Iodine	I	126.9044	Titanium	Ti	47.90
Iridium	Ir	192.2	Tungsten	W	183.85
Iron	Fe	55.847	Unnilquadium	Unq	(261.11)

Table 1.9. (Continued)

Krypton	Kr	83.80	Unnilpentium	Unp	(62.114)
Lanthanum	La	183.91	Unnilhexium	Unh	(263.118)
Lawrencium	Lr	(262.11)	Unnilseptium	Uns	(262.12)
Lead	Pb	207.19	Uranium	U	238.03
Lithium	Li	6.939	Vanadium	V	50.942
Lutetium	Lu	174.97	Xenon	Xe	131.30
Magnesium	Mg	24.312	Ytterbium	Yb	173.04
Manganese	Mn	54.9380	Yttrium	Y	88.905
Mendelevium	Md	(258.10)	Zinc	Zn	65.37
Mercury	Hg	200.59	Zirconium	Zr	91.22
Molybdenum	Mo	95.94			

1.2   Chemical Bonding

1.2.1   Covalent Bonding

A **covalent bond** is a bond in which a pair of electrons is shared between two atoms. Depending on the atoms electronegativity the bond is either polar or non-polar.

A pair of atoms with the same electronegativity would form a **non-polar covalent bond**, such as:

$$H\cdot \; + \; \cdot H \longrightarrow H\!:\!H$$

A **polar covalent bond** is one in which the atoms have different electronegativities, such as:

$$H\cdot \; + \cdot \overset{\cdot\cdot}{\underset{\cdot\cdot}{Cl}}\!\!: \; \longrightarrow \; H - \overline{Cl}\,|$$

1.2.2   Coordinate Covalent Bond (Dative Bond)

A **coordinate covalent bond** is a bond in which both pairs of electrons are donated by one atom and are shared between the two, for example:

1.2.3   Ionic Bonding

An **ionic bond** is one in which one or more electrons are transferred from one atom's valence shell (becoming a positively charged ion, called a **cation**) to the others valence shell (becoming a negatively charged ion, called a **anion**). The resulting electrostatic attraction between oppositely charged ions results in the formation of the ionic bond.

$$\overset{..}{Li} \cdot + \cdot \overset{..}{\underset{..}{F}} \mathbf{:} \longrightarrow Li^{+} + \mathbf{:} \overset{..}{\underset{..}{F}} \mathbf{:}$$

Not all compounds will be either purely covalent or purely ionic, most are somewhere in between. As a rule of thumb, if a compound has less than 50% ionic character it is considered covalent and more than 50% , ionic. The **ionic character** can be related to the difference in electronegativities of the bonded atoms. If the electronegativity difference is 1.7, the bond is about 50% ionic.

1.2.4   Dipole-Dipole Bonding

A **dipole-dipole bond** occurs between polar molecules and is a weak electrostatic attraction.

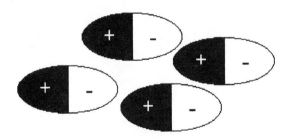

Figure 1.6. Dipole-dipole attraction

1.2.5   Ion-Dipole Bonding (Solvation)

An **ion-dipole bond** is another electrostatic attraction between an ion and several polar molecules. When an ionic substance is dissolved in a polar solvent, it is this kind of interaction that takes place. The negative ends of the solvent aligned themselves to the positive charge, and the positive ends aligned with the negative charge. This process is **solvation**. When the solvent is water the process is the same but called **hydration**.

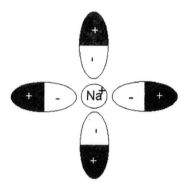

Figure 1.7. Ion-dipole attraction

1.2.6   Hydrogen Bonding

When hydrogen is bonded covalently to a small electronegative atom the electron cloud around the hydrogen is drawn to the electronegative atom and a strong dipole is created. The positive end of the dipole approaches close to the negative end of the neighboring dipole and a uniquely strong dipole-dipole bond forms, this is referred to as a **hydrogen bond**.

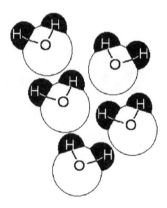

Figure 1.8. Hydrogen bonding

1.2.7   van der Waals

When the electron cloud around an atom or molecule shifts (for whatever reason), a temporary dipole is created, this in turns creates an induced dipole in the next molecule. This induced dipole (**van der Waals**) induces another and so on. The induced dipoles now are electrostaticly attracted to each other and a weak induced dipole attraction occurs.

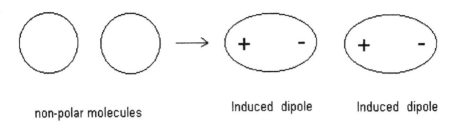

non-polar molecules　　　Induced dipole　　　Induced dipole

Figure 1.9. van der Waals attraction

1.3　Gases

The following laws and equations are for ideal gases. An **ideal gas** is considered to be composed of small particles whose volume is negligible when compared to the whole volume, and the gas particles do not exert any force on one another. Unless otherwise noted:

P - pressure
V - volume
T -absolute temperature (Kelvin)
n - number of moles
R - gas constant
　　8.314 J / mole K

1.3.1　Boyle's Law

For a fixed amount of gas, held at constant temperature, the volume is inversely proportional to the applied pressure.

$$PV = \text{constant}$$

1.3.2　Charles' Law

For a fixed amount of gas, held at constant pressure, the volume is directly proportional to the temperature.

$$V/T = \text{constant}$$

1.3.3　Gay-Lussac's Law

For a fixed amount of gas, held at constant volume, the pressure is directly proportional to the absolute temperature.

$$P/T = \text{constant}$$

1.3.4   Avogadro's Law

At constant temperature and pressure, equal volumes of gas contain equal amounts of molecules.

$$V/n = \text{constant}$$

1.3.5   Dalton's Law of Partial Pressures

The total pressure exerted by a mixture of several gases is equal to the sum of the gases individual pressure (partial pressure).

$$P_T = p_a + p_b + p_c + ...$$

P_T - pressure of gas a + b + c
p_a - pressure of gas a
p_b - pressure of gas b
p_c - pressure of gas c

1.3.6   Ideal Gas Law

Combining the above relationships and Avogadro's principle (under constant pressure and temperature, equal volumes of gas contain the equal numbers of molecules) into one equation we obtain the Ideal Gas Law:

$$PV = nRT$$

1.3.7   Equation of State of Real Gases

An ideal gas has negligible volume and exerts no force. However, real gases do have volumes and do exert forces upon one another. When these factors are taken into consideration, the following equation can be obtained:

$$(P + n^2a/V^2) - (V - nb) = nRT$$

a - proportionality constant
b - covolume

Note that a and b are dependent on the individual gas, since molecular volumes and molecular attractions vary from gas to gas.

Table 1.10. van der Waals constants for real gases.

Gas	a (atm liter^2 / mole^2)	b (liter / mole)
He	0.034	0.0237
O_2	1.36	0.0318
NH_3	4.17	0.0371
H_2O	5.46	0.0305
CH_4	2.25	0.0428

1.3.8 Changes of Pressure, Volume, or Temperature

By combining the equations for Boyle's law, Charles' law, and Gay-Lussac's law, a single equation can be obtained that is useful for many computations:

$$\frac{P_1 V_1}{T_1} = \frac{P_2 V_2}{T_2}$$

1.4 Solutions

1.4.1 Mass Percent

The **mass percent** of a solution is the mass of the solute divided by the total mass (solute + solvent) multiplied by 100.

$$\text{percent by mass of solute} = \frac{\text{mass of solute}}{\text{mass of solute} + \text{mass of solvent}} \times 100$$

1.4.2 Mole Fraction (*X*)

The **mole fraction (*X*)** is the number of moles of component A divided by the total number of moles of all components.

$$\text{mole fraction of A} = \frac{\text{moles of A}}{\text{moles of all components}}$$

1.4.3 Molarity (*M*)

The **molarity (*M*)** of a solution is the number of moles of solute dissolved in 1 liter of solvent.

$$\text{molarity} = \frac{\text{moles of solute}}{\text{liters of solvent}}$$

1.4.4   Molality (*m*)

Molality (*m*) is the number of moles of solute dissolved in 1000 g (1 kg) of solvent.

$$\text{molality} = \frac{\text{moles of solute}}{\text{mass of solvent}}$$

1.4.5   Dilutions

A handy and useful formula when calculating dilutions is:

$$M_{initial}\ V_{initial} = M_{final}\ V_{final}$$

1.5   Acids and Bases

1.5.1   Arrhenius Concept

An acid is any species that increases the concentration of **hydronium ions (H_3O^+)**, in aqueous solution.

A base is any species that increases the concentration of **hydroxide ion, (OH⁻)**, in aqueous solution.

For an acid:

$$HCl + H_2O \rightarrow H_3O^+ + Cl^-$$

For a base:

$$NH_3 + H_2O \rightarrow NH_4^+ + OH^-$$

However, the drawback with the Arrhenius concept is that it only applies to aqueous solutions.

1.5.2   Bronsted-Lowery Concept

An acid is a species which can donate a proton (i.e., a hydrogen ion, H^+) to a proton acceptor.

A base is a species which can accept a proton from a proton donor.

Along with the Bronsted-Lowery concept of a proton donor (acid) and a proton acceptor (base), arises the concept of **conjugate acid-base pairs**. For example, when the acid HCl reacts, it donates a proton thereby leaving Cl⁻ (which is now a proton acceptor, or the conjugate base of HCl). Using NH_3 as the base and H_2O as the acid:

$$\begin{array}{c} \quad\quad\quad \text{acid} \quad\quad\quad \text{conjugate base}\\ \quad\quad\quad \downarrow \quad\quad\quad\quad\quad \downarrow\\ NH_3 + H_2O \leftrightarrow NH_4^+ + OH^-\\ \uparrow \quad\quad\quad\quad\quad \uparrow\\ \text{base} \quad\quad\quad \text{conjugate acid} \end{array}$$

1.5.3   Lewis Concept

An acid is a species that can accept a pair of electrons.
A base is a species that can donate a pair of electrons.

1.6   Thermodynamics

1.6.1   First Law of Thermodynamics

The energy change of a system is equal to the heat absorbed by the system plus the work done by the system. The reason for the minus sign for work, w, is that any work done by the system results in a loss of energy for the system as a whole.

$$\Delta E = q - w$$

E = internal energy of the system
q = heat absorbed by the system
w = work done by the system

Table 1.11. Thermodynamic Processes

Process	Sign
work done by system	-
work done on system	+
heat absorbed by system (endothermic)	+
heat absorbed by surroundings (exothermic)	-

Making a substitution for work , the equation can be expressed as:

$$\Delta E = q - P\Delta V$$

For constant volume, the equation becomes:

$$\Delta E = q_v$$

1.6.2   Enthalpy

Enthalpy, H, is the heat content of the system at constant pressure.

$$\Delta E = q_p - P\Delta V$$

$$q_p = \Delta E + P\Delta V$$

$$q_p = (E_2 - E_1) + P(V_2 - V_1)$$

$$= (E_2 + PV_2) - (E_1 + PV_1)$$

$$q_p = H_2 - H_1 = \Delta H$$

1.6.3   Entropy

Entropy, S, is the measure of the degree of randomness of a system.

$$\Delta S = \frac{q_{rev}}{T}$$

T = temperature in °K

1.6.4   Gibbs Free Energy

Gibbs free energy, G, is the amount of energy available to the system to do useful work.

$$\Delta G = \Delta H - T\Delta S$$

$$\Delta G < 0 \quad \text{spontaneous process from } 1 \rightarrow 2$$
$$\Delta G > 0 \quad \text{spontaneous process from } 2 \rightarrow 1$$
$$\Delta G = 0 \quad \text{equilibrium}$$

1.6.5   Standard States

The **standard state** is the standard or normal condition of a species.

Table 1.12. Standard States

State of Matter	Standard State
Gas	1 atm pressure
Liquid	Pure liquid
Solid	Pure solid
Element	Free energy of formation = 0
Solution	1 molar concentration

Note also that ΔH°_f for an element in its natural state at 25°C and 1 atm is taken to be equal to zero.

1.6.6   Hess' Law of Heat Summation

The final value of ΔH for the overall process is the sum of all the enthalpy changes.

$$\Delta H^\circ = \Sigma \, \Delta H^\circ_f \,(\text{products}) - \Sigma \, \Delta H^\circ_f \,(\text{reactants})$$

For example, to vaporize 1 mole of H_2O at 100°C and 1 atm, the process absorbs 41 kJ of heat, $\Delta H = +41$ kJ.

$$H_2O\,(l) \;\rightarrow\; H_2O\,(g) \qquad \Delta H = +41 \text{ kJ}$$

If a different path to the formation of 1 mole of gasous H_2O is taken, the same amount of net heat will still be absorbed.

$$H_2(g) + \tfrac{1}{2}O_2\,(g) \;\rightarrow\; H_2O\,(l) \quad \Delta H_f = -283 \text{ kJ mol}^{-1}$$

$$H_2(g) + \tfrac{1}{2}O_2\,(g) \;\rightarrow\; H_2O\,(g) \quad \Delta H_f = -242 \text{ kJ mol}^{-1}$$

Reversing the first reaction, then adding the two reactions together and cancelling common terms, results in the original reaction, and the amount of heat absorbed by the system

$$H_2O \, (l) \; \rightarrow \; H_2 \, (g) + \tfrac{1}{2} \, O_2 \, (g) \quad \Delta H_f = +283 \text{ kJ mol}^{-1}$$

$$\underline{H_2 \, (g) + \tfrac{1}{2} \, O_2 \, (g) \; \rightarrow \; H_2O \, (g) \quad \Delta H_f = -242 \text{ kJ mol}^{-1}}$$

$$H_2O \, (l) \; \rightarrow \; H_2O \, (g) \qquad \Delta H = +41 \text{ kJ}$$

Using the Hess' Law of Summation:

$$\Delta H^\circ = \Sigma \, \Delta H^\circ_f \,(\text{products}) - \Sigma \, \Delta H^\circ_f \,(\text{reactants})$$
$$\Delta H^\circ = (-242 \text{ kJ}) - (-283 \text{ kJ}) = +41 \text{ kJ}$$

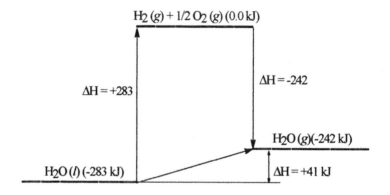

Figure 1.10. Enthalpy Diagram for $H_2O \, (l) \; \rightarrow \; H_2O \, (g)$

1.7   Equilibria

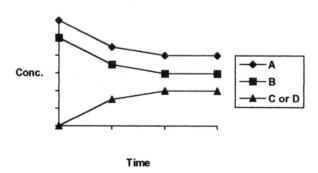

Figure 1.11. Concentrations of reactants and products approaching equilibrium.

When equilibrium is reached in a chemical reaction the rate of the forward reaction is equal to the rate of the reverse reaction, and the concentrations of the reactants and products do not change over time.

1.7.1   Homogeneous Equilibrium

Homogeneous equilibrium occurs when all reacting species are in the same phase. For the general reaction,

$$aA + bB \leftrightarrows cC + dD$$

the equation expressing the **law of mass action,** at equilibrium, is:

$$\frac{[C]^c[D]^d}{[A]^a[B]^b} = K_c$$

The quantity, K_c, is a constant, called the **equilibrium constant** (in this case it denotes the equilibrium constant for species in solution, expressed as moles per liter). The magnitude of K_c tells us to what extent the reaction proceeds. A large K_c indicates that the reactions proceeds to the right of the reaction. A low value indicates that the reaction proceeds to the right of the reaction.

For gas-phase equilibrium the expression becomes

$$K_p = \frac{P_C^c P_D^d}{P_A^a P_B^b}$$

where P is the partial pressures of the species in the reaction. K_p can be related to K_c by the following equation,

$$K_p = K_c(0.08206T)^{\Delta n}$$

T = the absolute temperature
Δn = moles of product - moles of reactants.

1.7.2   Heterogeneous Equilibrium

Heterogeneous equilibrium involves reactants and products in different phases. For example, when calcium carbonate is heated in a closed vessel, the following equilibrium reaction occurs:

$$CaCO_3(s) \rightleftarrows CaO(s) + CO_2(g)$$

The reactant is a solid, while the products are in the solid and gas phase. The equilibrium expression is written as the following:

$$K_c^{'} = \frac{[CaO][CO_2]}{[Ca\,CO_3]}$$

In any reaction that includes a solid, the solid concentration will remain constant and therefore is not included in the equilibrium expression. The equilibrium expression now becomes:

$$K_c^{''} \frac{[CaCO_3]}{CaO} = [CO_2] = K_c$$

1.7.3   Le Chatelier's Principle

Le Chatelier's Principle states that *when a system is in equilibrium and there is a change in one of the factors which affect the equilibrium, the system reacts in such a way as to cancel out the change and restore equilibrium.*.

An increase in temperature will shift the reaction in the direction of heat absorption.

An increase in the pressure will shift the reaction in the direction in which the number of moles is decreased.

An increase or decrease in pressure does not affect a reaction in which there is no variation in the number of moles.

An increase in the concentration of one of the components will cause the reaction to shift so as to decrease the added component.

1.7.4   Equilibrium of Water

The reaction for autoionization of water is:

$$H_2O + H_2O \rightleftarrows H_3O^+ + OH^-$$

The equilibrium expression is:

$$K = \frac{[H_3O^+][OH^-]}{[H_2O][H_2O]}$$

Since the concentration of water is a constant (≈ 55.55 M), $[H_2O]^2$ can be included in the equilibrium constant, K. this new constant is now called K_w.

$$K_w = K[H_2O]^2 = [H_3O^+]\,[OH^-]$$

K_w is the **ion product constant** for water, also called the **ionization constant** or **dissociation constant** for water. The ionization constant for water at 25 °C has a value of 1.0×10^{-14}. The equilibrium expression now becomes:

$$K_w = 1.0 \times 10^{-14} = [H_3O^+]\,[OH^-]$$

$$\text{Since } [H_3O^+] = [OH^-]$$

$$[H_3O^+] = [OH^-] = 1.0 \times 10^{-7}$$

When the concentration of hydrogen ions equals the concentration of hydroxide ions the solution is said to be neutral.

1.7.5   pH

pH is the measure of how strong or weak an acid is, and is defined as the negative of the log of the hydrogen ion concentration, or

$$pH = -\log[H_3O^+]$$

Water has a pH of 7, this is calculated from the dissociation constant for water:

$$[H_3O^+] = 1.0 \times 10^{-7}$$

$$pH = -\log[H_3O^+] = -\log(1.0 \times 10^{-7})$$

$$pH = 7$$

The concept of pH can be applied to any system in which hydrogen ions are produced. An acidic solution would have an excess of hydrogen ions, a basic solution would have an excess of hydroxide ions, and a neutral solution the hydrogen ions would equal the hydroxide ions. Since pH is a measure of the hydrogen ion concentration, acidic and basic solutions can be distinguished on the basis of their pH.

acidic solutions: $[H_3O^+] > 10^{-7}$ M, pH < 7
basic solutions: $[H_3O^+] < 10^{-7}$ M, pH > 7
neutral solutions: $[H_3O^+] = 10^{-7}$ M, pH = 7

1.7.6   Ionic Equilibrium

For a monoprotic acid HA, the equilibrium reaction is:

$$HA(aq) + H_2O \leftrightarrows H_3O^+(aq) + A^-(aq)$$

and the equilibrium expression is:

$$K_a = \frac{[H_3O^+][A^-]}{[HA]}$$

The equilibrium constant, K_a, is called the **acid dissociation constant.**
Similarly for a polyprotic acid (i.e. phosphoric acid), the equilibrium
reactions are:

$$H_3PO_4 \leftrightarrows H^+ + H_2PO_3 \qquad\qquad K_a' = \frac{[H^+][H_3PO_4^-]}{H_3PO_4} = 7.5\times10^{-3}$$

$$H_2PO_4^- \leftrightarrows H^+ + HPO_4^{2-} \qquad\qquad K_a'' = \frac{[H^+][HPO_4^{2-}]}{[H_2PO_4^-]} = 6.2\times10^{-8}$$

$$HPO_4^{2-} \leftrightarrows H^+ + PO_4^{3-} \qquad\qquad K_a''' = \frac{[H^+][PO_4^{3-}]}{[HPO_4^{2-}]} = 4.8\times10^{-13}$$

For a base the equilibrium reaction is:

$$B + H_2O \leftrightarrows BH^+ + OH^-$$

and the equilibrium expression is:

$$K_b = \frac{[BH^+][OH^-]}{[B]}$$

The equilibrium constant, K_b, is called the **base dissociation constant.**

1.7.7   Relationship between K_a and K_b Conjugate Pair

$$HA + H_2O \leftrightarrows H_3O^+ + A^-$$
$$\text{acid} \qquad\qquad\qquad \text{base}$$
$$\llcorner \quad \text{conjugates} \quad \lrcorner$$

$$K_a = \frac{[H_3O^+][A^-]}{[HA]}$$

$$A^- + H_2O \leftrightarrows HA + OH^-$$
$$\llcorner \text{ conjugates } \lrcorner$$

$$K_b = \frac{[HA][OH^-]}{[A^-]}$$

$$K_a K_b = [H_3O^+][OH^-] = 10^{-14}$$

$$pK_a + pK_b = 14$$

1.7.8   Hydrolysis

Hydrolysis is the between water and the ions of a salt.

1.7.8.1   Salt of a strong acid - strong base.

Consider NaCl, the salt of a strong acid and a strong base. The hydrolysis of this salt would yield NaOH and HCl. Since both species would completely dissociate into their respective ions yielding equivalent amounts of H_3O^+ and OH^-, the overall net effect would be that no hydrolysis takes place. Since $[H_3O^+] = [OH^-]$, the pH would be 7, a neutral solution.

1.7.8.2   Salt of a strong acid - weak base.

Consider the hydrolysis of NH_4Cl:

$$NH_4^+ + H_2O \leftrightarrows H_3O^+ + NH_3$$

$$K_h = \frac{[H_3O^+][NH_3]}{[NH_4^+]}$$

$$K_h = \frac{K_w}{K_b} = \frac{[H_3O^+][OH^-]}{[NH_4^+][OH^-]/[NH_3]} = \frac{[H_3O^+][NH_3]}{[NH_4^+]}$$

1.7.8.3   Salt of a weak acid - strong base.

Consider the hydrolysis of $NaC_2H_3O_2$:

$$C_2H_3O_2^- + H_2O \rightleftharpoons HC_2H_3O_2 + OH^-$$

$$K_h = \frac{[HC_2H_3O_2][OH^-]}{[C_2H_3O_2^-]}$$

$$K_h = \frac{K_w}{K_a} = \frac{[H_3O^+][OH^-]}{[H_3O^+][C_2H_3O_2^-]/[HC_2H_3O_2]} = \frac{[HC_2H_3O_2][OH^-]}{[C_2H_3O_2^-]}$$

1.7.9   Solubility Product

In the case for which a solid is being dissolved, the general chemical reaction becomes:

$$A_aB_b \rightleftharpoons aA + bB$$

and the equilibrium expression is:

$$K = \frac{[A]^a[B]^b}{[A_aB_b]}$$

the denominator in the expression $[A_aB_b]$ represents the concentration of the pure solid and is constant, therefore it can be incorporated into the equilibrium constant, K. The expression now becomes:

$$K_{sp} = [A]^a[B]^b$$

For example, a saturated solution of AgCl, woulld have the following equilibrium:

$$AgCl\,(s) \leftrightarrows Ag^+ + Cl^-$$

$$K_{sp} = [Ag^+]\,[Cl^-]$$

The value of the K_{sp} for AgCl is 1.7×10^{-10}

$$1.7 \times 10^{-10} = [Ag^+]\,[Cl^-]$$

If the ion product is equal to or less that the K_{sp} no precipitate will form. If the ion product is greater than the K_{sp} value, the material will precipitate out of solution so that the ion product will be equal to the K_{sp}.

Table 1.13. Solubility Products

Compound	K_{sp}
AgCl	1.7×10^{-10}
AgBr	5.0×10^{-13}
$BaSO_4$	1.5×10^{-9}
CuS	8.5×10^{-36}
$PbCl_2$	1.6×10^{-5}
HgS	1.6×10^{-54}

1.7.10   Common Ion Effect

The **common ion** is when an ion common to one of the salt ions is introduced to the solution. The introduction of a common ion produces an effect on the equilibrium of the solution and according to Le Chatelier's principle, i.e. the equilibrium is shifted so as to reduce the effect of the added ion. This is referred to as the **common ion effect**.

In the case of a solution of AgCl, if NaCl is added, the common ion being Cl^-, the equilibrium would be shifted to the left so that the ion product will preserve the value of the K_{sp}.

1.8   Kinetics

Kinetics deals with the rate (how fast) that a chemical reaction proceeds with. The reaction rate can be determined by following the concentration of either the reactants or products. The rate is also dependent on the concentrations, temperature, catalysts, and nature of reactants and products.

1.8.1   Zero-Order Reactions

Zero-order reactions are independent of the concentrations of reactants.

$$A \rightarrow B$$

$$\text{rate} = -\frac{\Delta[A]}{\Delta t} = k[A]^0 = k$$

1.8.2   First-Order Reactions

First-order reactions are dependent on the concentration of the reactant.

$$A \rightarrow B$$

$$\text{rate} = k[A]^1 = k[A]$$

1.8.3   Second-Order Reactions

There are two type of second-order reactions. The first kind involves a single kind of reactant.

$$2A \rightarrow B$$

$$\text{rate} = k[A]^2$$

The second kind of reaction involves two different kind of reactants.

$$A + B \rightarrow C$$

$$\text{rate} = k[A][B]$$

1.8.4   Collision Theory

Consider the decomposition of HI.

$$2HI\,(g) \rightarrow H_2\,(g) + I_2\,(g)$$

In order for the decomposition of HI to take place, two molecules of HI must collide with each other with the proper orientation as shown in Figure 1.12. If the molecules collide without the proper orientation then no decomposition takes place.

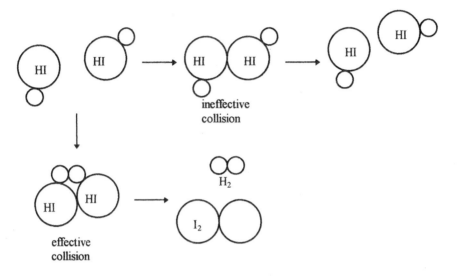

Figure 1.12. Effective and Ineffective collisions

Not all collisions with the proper orientation will react. Only those collisions with the proper orientation and sufficient energy to allow for the breaking and forming of bonds will react. The minimum energy available in a collision which will allow a reaction to occur is called the **activation energy**.

1.8.5 Transition State Theory

When a collision with the proper orientation and sufficent activation energy occurs, an intermediate state exists before the products are formed. This intermediate state, also called an **activated complex** or **transition state**, is neither the reactant or product, but rather a highly unstable combination of both, as represented in Figure 1.13 for the decomposition of HI.

$$\begin{matrix} H \\ | \\ I \end{matrix} + \begin{matrix} H \\ | \\ I \end{matrix} \rightleftharpoons \left[\begin{matrix} H\text{-}\text{-}H \\ | \quad | \\ I\text{-}\text{-}\text{-}I \end{matrix} \right] \longrightarrow \begin{matrix} H\text{-}\text{-}H \\ + \\ I\text{—}I \end{matrix}$$

Figure 1.13. Transition state or activated complex

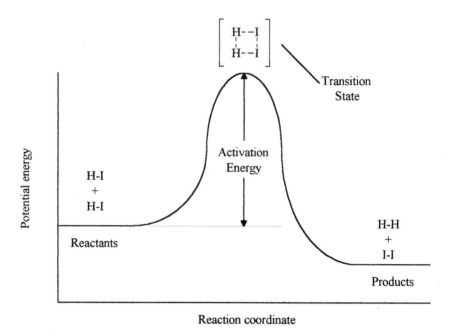

Figure 1.14. Potential energy diagram for the decomposition of HI.

Figure 1.14 shows the potential energy diagram for the decomposition of HI. As can be seen, in order to reach the activated complex or transition state the proper orientaion and sufficent collision energy must be achieved. Once these requirements are achieved the reaction continues on to completion and the products are formed.

1.8.6   Catalysts

A **catalyst** is a substance that affects the rate of a chemical reaction without itself being consumed or chemically altered. The catalyst takes part in the reaction by providing an alternative route to the production of products. The catalyzed reaction has a lower activation energy than that of the uncatalyzed reaction, as shown in Figure 1.15. By lowering the activation energy there are more molecules with sufficent energy that can react and thus the rate of the reaction is affected.

A **homogeneous catalyst** is in the same phase as the reactants. The catalyst and the reactants form a reactive intermediate.

A **heterogeneous catalyst** is not in the same phase as the reactants. The reactants are absorbed on the surface of the heterogeneous catalyst and the reaction then takes place.

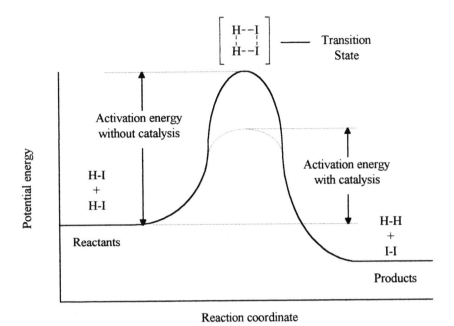

Figure 1.15. Energy diagram for a reaction with and without a catalysts

Chapter 2

Inorganic Chemistry

2.1   Group IA Elements

Alkali Metals -Li, Na, K, Rb, Cs, Fr

Table 2.1. Group IA Properties

Element	Li	Na	K	Rb	Cs	Fr
Electronic configuration	[He]2s	[Ne]3s	[Ar]4s	[Kr]5s	[Xe]6s	[Rn]7s
M.P. (^{o}K)	453.7	371.0	336.35	312.64	301.55	300
B.P. (^{o}K)	1615	1156	1032	961	944	950
Pauling's Electronegativity	0.98	0.93	0.82	0.82	0.79	0.7
Atomic radius (Å)	2.05	2.23	2.77	2.98	3.34	-
Covalent radius (Å)	1.23	1.54	2.03	2.16	2.35	-
Ionic radius (Å)(+1)	0.68	0.98	1.33	1.48	1.67	1.8
Ionization enthalpy (eV)	5.392	5.139	4.341	4.177	3.894	-
Crystal Structure	bcc	bcc	bcc	bcc	bcc	bcc

Table 2.2. Group IA Compounds

	Li	Na	K	Rb	Cs	Fr
H^-	x	x	x	x	x	-
X^-	x	x	x	x	x	-
CH_3COO^-	x	x	x	x	x	-
HCO_3^-	x	x	x	x	x	-
ClO^-	-	x	x	-	-	-
ClO_3^-	x	x	x	x	x	-
ClO_4^-	x	x	x	x	x	-
OH^-	x	x	x	x	x	-
NO_3^-	x	x	x	x	x	-
NO_2^-	x	x	x	-	x	-
$H_2PO_4^-$	x	x	x	-	-	-
HSO_4^-	x	x	x	x	x	-
HSO_3^-	-	x	x	-	-	-
CO_3^{-2}	x	x	x	x	x	-
$C_2O_4^{-2}$	x	x	x	-	x	-
HPO_4^{-2}	-	x	x	-	-	-
SO_4^{-2}	x	x	x	x	x	-
SO_3^{-2}	x	x	x	-	-	-
PO_3^{-3}	x	x	x	-	-	-
PO_4^{-3}	x	x	x	-	-	-
N^{-3}	x	x	x	-	-	-

2.2   Group IIA Elements

Alkaline Earth Metals -Be, Mg, Ca, Sr, Ba, Ra

Table 2.3. Group IIA Properties

Element	Be	Mg	Ca	Sr	Ba	Ra
Electronic configuration	$[He]2s^2$	$[Ne]3s^2$	$[Ar]4s^2$	$[Kr]5s^2$	$[Xe]6s^2$	$[Rn]7s^2$
M.P. (^oK)	1560	922	1112	1041	1002	973
B.P. (^oK)	2745	1363	1757	1650	2171	1809
Pauling's Electronegativity	1.57	1.31	1.00	0.95	0.89	0.9
Atomic radius (Å)	1.40	1.72	2.223	2.45	2.78	-
Covalent radius (Å)	0.90	1.36	1.74	1.91	1.98	-
Ionic radius (Å)(+2)	0.35	0.66	1.18	1.112	1.34	1.43
Ionization enthalpy (eV)	9.322	7.646	6.113	5.695	5.212	5.279
Crystal Structure	hex	hex	fcc	fcc	bcc	bcc

Table 2.4. Group IIA Compounds

	Be	Mg	Ca	Sr	Ba	Ra
H^-	x	x	x	x	x	-
X^-	x	x	x	x	x	-
CH_3COO^-	x	x	x	x	x	-
HCO_3^-	-	-	-	-	-	-
ClO^-	-	-	x	-	x	-
ClO_3^-	-	x	x	x	x	-
ClO_4^-	-	x	x	x	x	-
OH^-	x	x	x	x	x	-
NO_3^-	x	x	x	x	x	-
NO_2^-	-	x	x	x	x	-
$H_2PO_4^-$	-	-	x	-	x	-
HSO_4^-	-	-	-	x	-	-
HSO_3^-	-	-	x	-	-	-
CO_3^{-2}	x	x	x	x	x	-
$C_2O_4^{-2}$	x	x	x	x	x	-
HPO_4^{-2}	-	x	x	x	x	-
SO_4^{-2}	x	x	x	x	x	-
SO_3^{-2}	-	x	x	x	x	-
PO_3^{-3}	-	-	x	-	x	-
PO_4^{-3}	x	x	x	-	x	-
N^{-3}	x	x	x	x	x	-

2.3   Group IIIA Elements

Boron group - B, Al, Ga, In, Tl

Table 2.5. Group IIIA Properties

Element	B	Al	Ga	In	Tl
Electronic configuration	[He]$2s^22p$	[Ne]$3s^23p$	[Ar]$3d^{10}$ $4s^24p$	[Kr]$4d^{10}$ $5s^25p$	[Xe]$4f^{14}$ $5d^{10}6s^26p$
M.P. (^oK)	2300	933.25	301.90	429.76	577
B.P. (^oK)	4275	2793	2478	2346	1746
Pauling's Electronegativity	2.04	1.61	1.81	1.78	2.04
Atomic radius (Å)	1.17	1.82	1.81	2.00	2.08
Covalent radius (Å)	0.82	1.18	1.26	1.44	1.48
Ionic radius (Å)(+3)	0.23	0.51	0.81	0.81	0.95
Ionization enthalpy (eV)	8.298	5.986	5.999	5.786	6.108
Crystal Structure	rhom	fcc	orthorho	tetrag	hex

Table 2.6. Group IIIA Compounds

	B	Al	Ga	In	Tl
H^-	x	-	x	-	-
X^-	x	x	x	x	x
CH_3COO^-	-	x	x	-	x
HCO_3^-	-	-	-	-	-
ClO^-	-	-	-	-	-
ClO_3^-	-	-	-	-	x
ClO_4^-	-	-	x	x	x
OH^-	-	x	x	x	x
NO_3^-	-	x	x	x	x
NO_2^-	-	-	-	-	x
$H_2PO_4^-$	-	-	-	-	x
HSO_4^-	-	-	-	-	x
HSO_3^-	-	-	-	-	x
CO_3^{-2}	-	-	-	-	x
$C_2O_4^{-2}$	-	x	x	-	x
HPO_4^{-2}	-	-	-	-	-
SO_4^{-2}	-	x	x	x	x
SO_3^{-2}	-	-	-	-	x
PO_3^{-3}	-	x	-	-	-
PO_4^{-3}	-	x	-	-	x
N^{-3}	x	x	x	-	-

2.4   Group IVA Elements

Carbon group - C, Si, Ge, Sn, Pb

Table 2.7. Group IVA Properties

Element	C	Si	Ge	Sn	Pb
Electronic configuration	$[He]2s^2$ $2p^2$	$[Ne]3s^2$ $3p^2$	$[Ar]3d^{10}$ $4s^24p^2$	$[Kr]4d^{10}$ $5s^25p^2$	$[Xe]4f^{14}$ $5d^{10}6s^2$ $6p^2$
M.P. (^{o}K)	4100	1685	1210.4	505.06	600.6
B.P. (^{o}K)	4470	3540	3107	2876	2023
Pauling's Electronegativity	2.55	1.90	2.01	1.96	2.33
Atomic radius (Å)	0.91	1.46	1.52	1.72	1.81
Covalent radius (Å)	0.77	1.11	1.22	1.41	1.47
Ionic radius (Å)(xx)					
Ionization enthalpy (eV)	11.260	8.151	7.899	7.344	7.416
Crystal Structure	hex	fcc	orthorho	tetrag	fcc

2.5   Group VA Elements

Nitrogen group - N, P, As, Sb, Bi

Table 2.8. Group VA Properties

Element	N	P	As	Sb	Bi
Electronic configuration	$[He]2s^2$ $2p^3$	$[Ne]3s^2$ $3p^3$	$[Ar]3d^{10}$ $4s^24p^3$	$[Kr]4d^{10}$ $5s^25p^3$	$[Xe]4f^{14}$ $5d^{10}6s^2$ $6p^3$
M.P. (^{o}K)	63.14	317.3	1081	904	544.52
B.P. (^{o}K)	77.35	550	876 (sub)	1860	1837
Pauling's Electronegativity	3.04	2.19	2.18	2.05	2.02
Atomic radius (Å)	0.75	1.23	1.33	1.53	1.63
Covalent radius (Å)	0.75	1.06	1.20	1.40	1.46
Ionic radius (Å)(xx)					
Ionization enthalpy (eV)	14.534	10.486	9.81	8.641	7.289
Crystal Structure	hex	monoclini	rhom	rhom	rhom

2.6   Group VIA Elements

Oxygen group - O, S, Se, Te, Po

Table 2.9. Group VIA Properties

Element	O	S	Se	Te	Po
Electronic configuration	$[He]2s^2$ $2p^4$	$[Ne]3s^2$ $3p^4$	$[Ar]\,3d^{10}$ $4s^24p^4$	$[Kr]4d^{10}$ $5s^25p^4$	$[Xe]4f^{14}$ $5d^{10}6s^2$ $6p^4$
M.P. ($^{\circ}$K)	50.35	388.36	494	722.65	527
B.P. ($^{\circ}$K)	90.18	717.75	958	1261	1235
Pauling's Electronegativity	3.44	2.58	2.55	2.1	2.0
Atomic radius (Å)	0.65	1.09	1.22	1.42	1.53
Covalent radius (Å)	0.73	1.02	1.16	1.36	1.46
Ionic radius (Å)(-2)	1.32	1.84	1.91	2.11	-
Ionization enthalpy (eV)	13.618	10.360	9.752	9.009	8.42
Crystal Structure	cubic	orthorho	hex	hex	monoclin

2.7   Group VIIA Elements

Halogens - F, Cl, Br, I, At

Table 2.10. Group VIIA Properties

Element	F	Cl	Br	I	At
Electronic configuration	$[He]2s^2$ $2p^5$	$[Ne]3s^2$ $3p^5$	$[Ar]\,3d^{10}$ $4s^24p^5$	$[Kr]4d^{10}$ $5s^25p^5$	$[Xe]4f^{14}$ $5d^{10}6s^2$ $6p^5$
M.P. ($^{\circ}$K)	53.48	172.16	265.90	386.7	575
B.P. ($^{\circ}$K)	84.95	239.1	332.25	458.4	610
Pauling's Electronegativity	3.98	3.16	2.96	2.66	2.2
Atomic radius (Å)	0.57	0.97	1.12	1.32	1.43
Covalent radius (Å)	0.72	0.99	1.14	1.33	1.45
Ionic radius (Å)(-1)	1.33	1.81	1.96	2.20	-
Ionization enthalpy (eV)	17.411	112.967	11.814	10.451	-
Crystal Structure	cubic	orthorho	orthorho	orthorho	-

2.8   Group VIIIA Elements

Noble (Inert) Gases - He, Ne, Ar, Kr, Xe, Rn

Table 2.11. Group VIIIA Properties

Element	He	Ne	Ar	Kr	Xe	Rn
Electronic configuration	$1s^2$	[He] $2s^22p^6$	[Ne] $3s^23p^6$	[Ar] $3d^{10}4s^2$ $4p^6$	[Kr] $4d^{10}5s^2$ $5p^6$	[Xe]$4f^{14}$ $5d^{10}6s^26p^6$
M.P. (^{o}K)	0.95	24.553	83.81	115.78	165.03	202
B.P. (^{o}K)	4.215	27.096	87.30	119.80	161.36	211
Pauling's Electronegativity	0	0	0	0	0	0
Atomic radius ($\mathring{A}$)	0.49	0.51	0.88	1.03	1.24	1.34
Covalent radius ($\mathring{A}$)	0.93	0.71	0.98	1.12	1.31	-
Ionic radius ($\mathring{A}$)	-	-	-	-	-	-
Ionization enthalpy (eV)	24.58	21.56	15.75	13.99	12.13	10.74
Crystal Structure	hex	fcc	fcc	fcc	fcc	fcc

Table 2.12. Some Group VIIIA Compounds

Oxidation State	Compound	Form	Mp (^{o}C)	Structure
II	XeF_2	Crystal	129	Linear
IV	XeF_4	Crystal	117	Square
VI	XeF_6	Crystal	49.6	
	Cs_2XeF_8	Solid		
	$XeOF_4$	Liquid	-46	Sq. Pyramid
	XeO_3	Crystal		Pyramidal
VIII	XeO_4	Gas		Tetrahedral
	XeO_6^{4-}	Salts		Octahedral

2.9 Transition Metal Elements

Table 2.13. Transition Elements Properties

Element	Elec. conf.	M.P. (°C)	B.P. (°C)	Density (g/cm³)	Atomic Radius	Ionic Radius (Å)			
						+2	+3	+4	+X
Sc	$3d^1 4s^2$	1540	2730	3.0	1.61		0.81		
Y	$4d^1 5s^2$	1500	2927	4.472	1.8		0.92		
La	$5d^1 6s^2$	920	3470	6.162	1.86		1.14		
Ti	$3d^2 4s^2$	1670	3260	4.5	1.45	0.90	0.87	0.68	
Zr	$4d^2 5s^2$	1850	3580	6.4	1.60			0.79	
Hf	$5d^2 6s^2$	2000	5400	13.2	-			0.78	
									X=5
V	$3d^3 4s^2$	1900	3450	5.8	1.32	0.88	0.74		0.59
Nb	$4d^4 5s^1$	2420	4930	8.57	1.45				0.69
Ta	$5d^3 6s^2$	3000	-	-	1.45				0.68
									X=6
Cr	$3d^5 4s^1$	1900	2640	7.2	1.37	0.88	0.63		0.52
Mo	$4d^5 5s^1$	2610	5560	10.2	1.36				0.62
W	$5d^4 6s^2$	3410	5930	19.3	1.37				0.62
									X=7
Mn	$3d^5 4s^2$	1250	2100	7.4	1.37	0.80	0.66		0.46
Tc	$4d^5 5s^2$	2140	-	-	1.36				
Re	$5d^5 6s^2$	3180	-	21	1.37				0.56
Fe	$3d^6 4s^2$	1540	3000	7.9	1.24	0.76	0.64		
Ru	$4d^7 5s^1$	2300	3900	12.2	1.33			0.67	
Os	$5d^6 6s^2$	3000	5500	22.4	1.34			0.69	
Co	$3d^7 4s^2$	1490	2900	8.9	1.25	0.74	0.63		
Rh	$4d^8 5s^1$	1970	3730	12.4	1.34	0.86	0.68		
Ir	$5d^7 6s^2$	2450	4500	22.5	1.36			0.68	
Ni	$3d^8 4s^2$	1450	2730	8.9	1.25	0.72	0.62		
Pd	$4d^{10}$	1550	3125	12.0	1.38	0.80		0.65	
Pt	$5d^9 6s^1$	1770	3825	21.4	1.39	0.80		0.65	
									X=1
Cu	$3d^{10} 4s^1$	1083	2600	8.9	1.28	0.69			0.96
Ag	$4d^{10} 5s^1$	961	2210	10.5	1.44	0.89			1.26
Au	$5d^{10} 6s^1$	1063	2970	19.3	1.44		0.85		1.37
Zn	$3d^{10} 4s^2$	419	906	7.3	1.33	0.74			
Cd	$4d^{10} 5s^2$	321	765	8.64	1.49	0.97			
Hg	$5d^{10} 6s^2$	-39	357	13.54	1.52	1.10			

Table 2.14. Oxidation States of Transition Elements

Sc	Ti	V	Cr	Mn	Fe	Co	Ni	Cu	Zn
+III	+III	+II	+II	+II	+II	+II	+II	+I	+II
	+IV	+III	+III	+III	+III	+III	+III	+II	
		+IV	+VI	+IV	+IV				
		+V		+VI	+VI				
				+VII					

	Ce	Mo			Ru	Rh	Pd	Ag	Cd
	+III	+III			+II	+III	+II	+I	+II
	+IV	+IV			+III	+IV	+IV	+II	
		+V			+IV				
		+VI			+VI				

		W			Os	Ir	Pt	Au	Hg
		+IV			+IV	+III	+II	+I	+I
		+V			+VI	+IV	+IV	+III	+II
		+VI			+VIII	+VI			

2.10   Ionic Solids

Ionic solids are mostly inorganic compounds, which are held together by ionic bonds (Chapter 1.2.3). The ions are treated as thought they were hard spheres with either positive (cations) or negative (anions) charges. Using this model the ionic solids or **crystals**, are found to be arranged in a very specific order called a **crystal lattice**. The crystal lattice consists of repeating units called unit cells (see below).

2.11   Coordination Number

For ionic compounds the **coordination number** is the number of anions that are arranged about the cation in a organized structure. For example, NaCl has a coordination number of 6. In otherwords, 6 Cl^- atoms surround 1 Na^+ atom. The number of anions that can surround a cation is dependent (but not entirely) on the relative sizes of the ions involved. Table 2.15 illustrates the ratios of the radii of the ions and their coordination number.

Table 2.15. Radius Ratios and Coordination Number

Coordination Number	Geometry	Ratio (+/-)
2	Linear	0.000 - 0.155
3	Trigonal	0.155 - 0.225
4	Tetrahedral	0.225 - 0.414
4/6	Sq. Planar/Octahedral	0.414 - 0.732
8	Cubic	0.732 - 1.000
12	Dodecahedral	1.000 -

Taking NaCl, the radii of Na^+ ion is 0.95Å and Cl^- is 1.81Å. Their ratio would be as follows:

$$\frac{r_{cation}}{r_{anion}} = \frac{r_{Na}}{r_{Cl}} = \frac{0.95}{1.81} = 0.52$$

forming the sodium chloride lattice with coordination number 6.

2.12   Ionic Crystal Systems

A crystal lattice can be broken down to a small repeating unit called a **unit cell**. There are only seven distinct unit cells possible, and are shown in Figure 2.1.

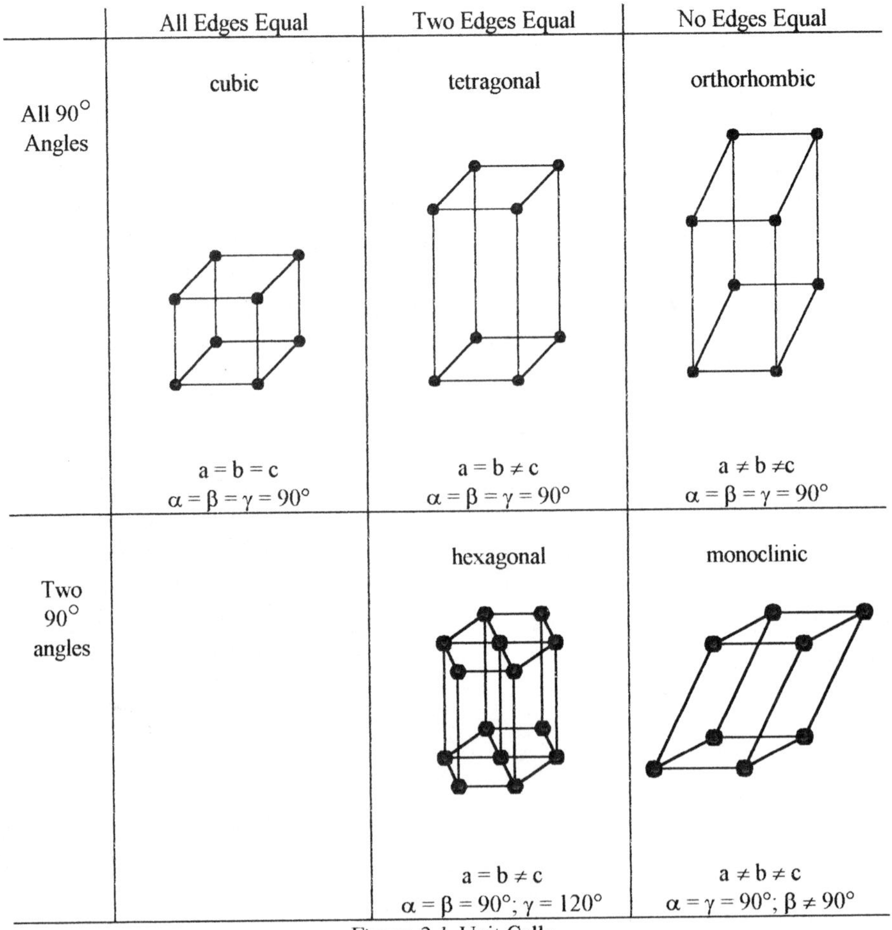

	All Edges Equal	Two Edges Equal	No Edges Equal
All 90° Angles	cubic a = b = c α = β = γ = 90°	tetragonal a = b ≠ c α = β = γ = 90°	orthorhombic a ≠ b ≠ c α = β = γ = 90°
Two 90° angles		hexagonal a = b ≠ c α = β = 90°; γ = 120°	monoclinic a ≠ b ≠ c α = γ = 90°; β ≠ 90°

Figure 2.1. Unit Cells

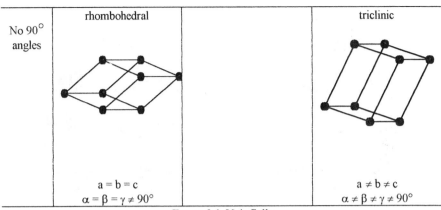

No 90° angles	rhombohedral		triclinic
	$a = b = c$ $\alpha = \beta = \gamma \neq 90°$		$a \neq b \neq c$ $\alpha \neq \beta \neq \gamma \neq 90°$

Figure 2.1. Unit Cells

2.13 Crystal Lattice Packing

When a crystal lattice forms the ions are arranged in the most efficient way of packing spheres into the smallest possible space. Starting with a single layer as in figure 2.2a, a second layer can be placed on top of it in the hollows of the first Figure 2.2b. At this point a third layer can be placed. If the third layer is placed directly over the first, Figure 2.2c, the structure is **hexagonal close-packed (hcp).** The layering in hexagonal close-packed is ABABAB. If the third layer is placed so it is not directly over the first a different arrangement is obtained. This structure is **cubic close-packed (ccp).** The layering in cubic close-packed is ABCABC.

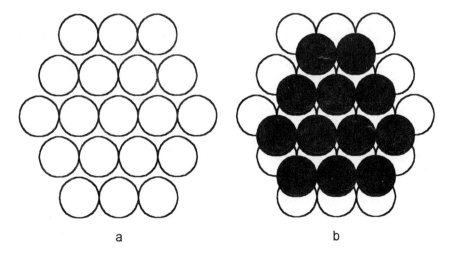

a   b

Figure 2.2. Close Packing of Atoms

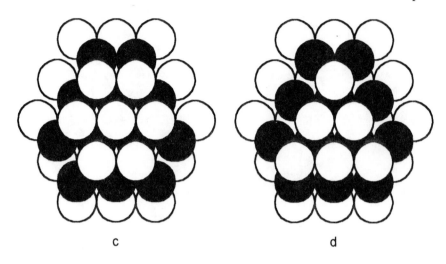

c   d

Figure 2.2. Close Packing of Atoms

2.14   Crystal Lattice Types

Shown below are commonly encountered crystal lattice structures. The lattice type depends on the radius ratio favoring a particular coordination number for the structure type.

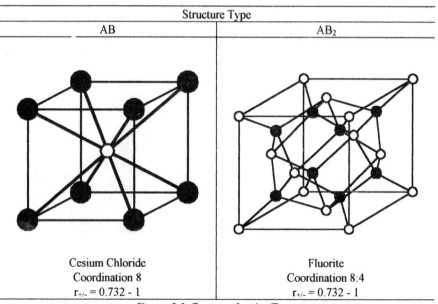

Structure Type	
AB	AB$_2$
Cesium Chloride Coordination 8 $r_{+/-} = 0.732 - 1$	Fluorite Coordination 8:4 $r_{+/-} = 0.732 - 1$

Figure 2.3. Common Lattice Types

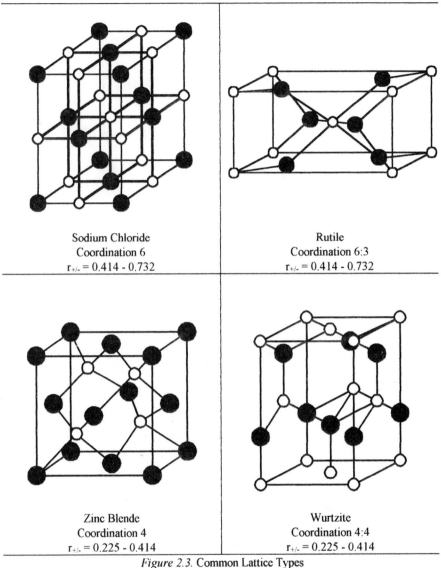

Figure 2.3. Common Lattice Types

2.15   Crystal Lattice Energy

2.15.1   Born-Haber Cycle

An important property of an ionic crystal is the energy required to break the crystal apart into individual ions, this is the **crystal lattice energy**. It can be measured by a thermodynamic cycle, called the **Born-Haber** cycle.

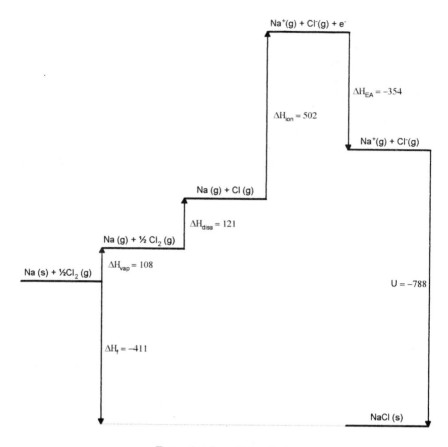

Figure 2.4. Born-Haber Cycle

The Born-Haber cycle follows the Law of Conservation of Energy, that is when a system goes through a series of changes and is returned to its initial state the sum of the energy changes is equal to zero. Thus the equation:

$$0 = \Delta h_f + \Delta h_{vap} + \tfrac{1}{2} \Delta h_{diss} + \Delta H_{ion} + \Delta H_{EA} + U$$

From this the crystal lattice energy, U, can be calculated from the following enthalpies:

enthalpy of formation	(Δh_f)	-411
vaporization of sodium	(Δh_{vap})	-108
dissociation of $Cl_2(g)$ into gaseous atoms	$(\tfrac{1}{2} \Delta h_{diss})$	-121
ionization of Na(g) to $Na^+(g)$	(Δh_{ion})	-502
electron attachment to Cl(g) to give $Cl^-(g)$	(ΔHea)	354
crystal lattice energy	U	-788kJ/mol

2.15.2 Madelung Constant

The crystal lattice energy can be estimated from a simple electrostatic model When this model is applied to an ionic crystal only the electrostatic charges and the shortest anion-cation internuclear distance need be considered. The summation of all the geometrical interactions between the ions is called the **Madelung constant**. From this model an equation for the crystal lattice energy is derived:

$$U = -1389 \frac{M}{r} \left(1 - \frac{1}{n}\right)$$

U = crystal lattice energy
M = Madelung constant
r = shortest internuclear distance
n = Born exponent

The Madelung constant is unique for each crystal structure and is defined only for those whose interatomic vectors are fixed by symmetry. The **Born exponent**, n, can be estimated for alkali halides by the noble-gas-like electron configuration of the ion. It can also be estimated from the compressibility of the crystal system. For NaCl, n equals 9.1.

Table 2.16. Madelung Constants		*Table 2.17.* Born Exponents	
Structure Type	M	Configuration	n
NaCl	1.74756	He	5
CsCl	1.76267	Ne	7
CaF$_2$	5.03878	Ar	9
Zinc Blende	1.63805	Kr	10
Wurtzite	1.64132	Xe	12

For NaCl, substituting in the appropriate values into the equation we obtain:

$$U = -1389 \frac{1.747}{2.82} \left(1 - \frac{1}{9.1}\right)$$

$$U = -860 + 95 = -765 \text{ kJ mol}^{-1}$$

As can be, seen the result is close (within 3%) of the value of U obtained from using the Born-Haber cycle. More accurate calculations can be obtained if other factors are taken into account, such as, van der Waals repulsion, zero-point energy, etc....

2.16   Complexes

Transition and non-transition metal ions form a great many complex ions and molecules. Bonding is achieved by an ion or molecule donating a pair of electrons to the metal ion. This type of bond is an coordinate covalent bond (section 1.2.2), the resulting complexes are called **coordination complexes.** The species donating the electron pair is called a **ligand.** More than one type of ligand can bond to the same metal ion, i.e. K_2PtCl_6. In addition a ligand can bond to more than one site on the metal ion, a phenomenon called **chelation**.

The bonding involved in the formation of coordination complexes involve the d orbitals (section 1.1.3) of the metal ion. The electron pair being donated occupies the empty d orbitals and accounts for the geometry of the complex.

2.16.1   Unidentate Ligands

CO, CN^-, NO_2^-, NH_3, SCN^-, H_2O, F^- RCO_2^-, OH^- Cl^-, Br^-, I^-

2.16.2   Bidentate Ligands

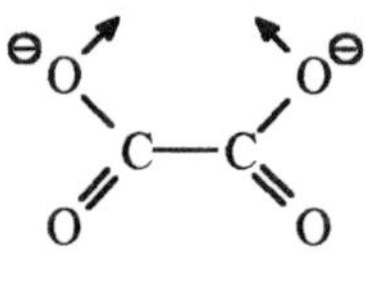

Oxalate Ion

Ethylenediamine

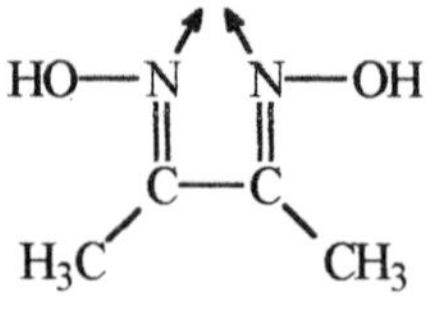

Diacetyldioxime

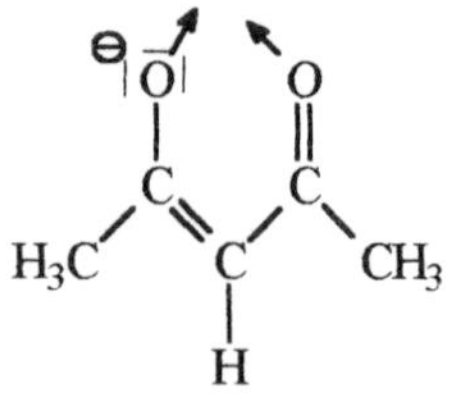

Acetylacetonate

2.16.3 Tridentate Ligands

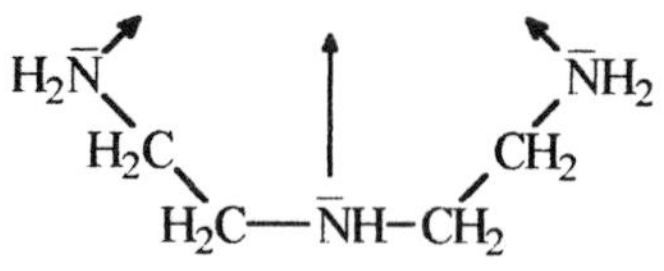

Diethylenetriamine

2.16.4 Quadridentate Ligands

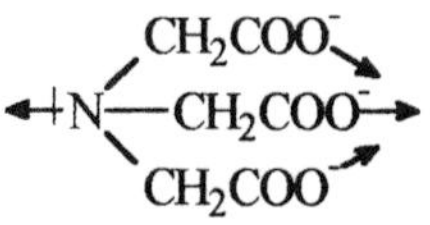

Tricarboxymethylamine

2.16.5 Pentadentate Ligands

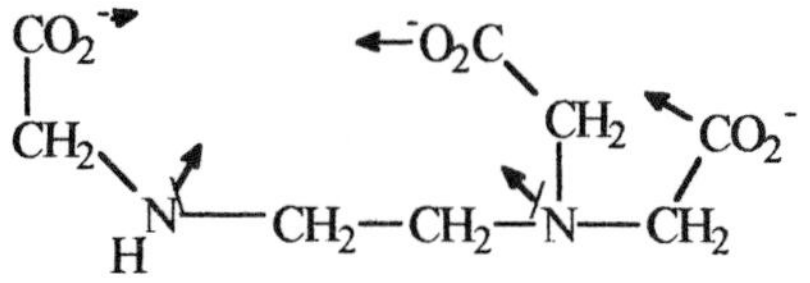

Ethylenediaminetriacetic acid

2.16.6 Hexadentate Ligands

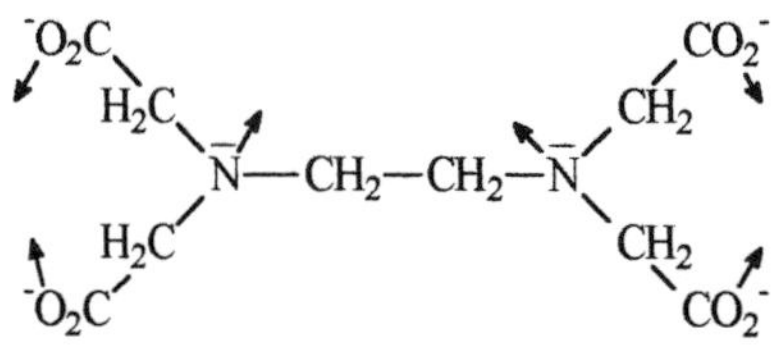

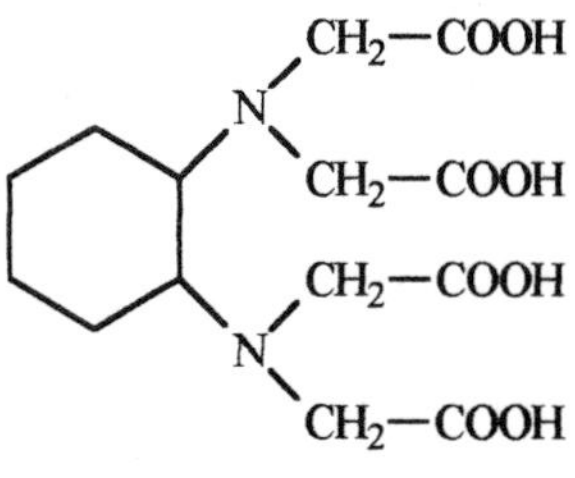

Ethylenediaminetetraacetic acid
- EDTA

Diaminocyclohexanetetraacetic
acid - CDTA

Diethylenetriaminepentaacetic
acid - DTPA

Dioxaoctamethylenedinitriolo
tetraacetic acid - PGTA

Chapter 3

Organic Chemistry

3.1   Classification Of Organic Compounds

3.1.1   General Classification

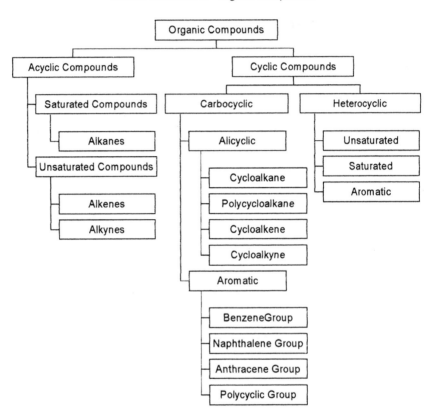

Figure 3.1. General Classification of Organic Compounds

3.1.2   Classification by Functional Group

Table 3.1. Organic Functional Groups

Type	Functional Group	Example	Name
Alkane	R-H	$CH_3CH_2CH_3$	propane
Alkene	R=R	$CH_2=CHCH_3$	propene
Diene	R=R-R=R	$CH_2=CH-CH=CH_2$	1,4-butene
Alkyne	R ≡ R	$CH \equiv CH$	ethyne
Halide	R-X	CH_3CH_2-Br	bromoethane
Alcohol	R-OH	CH_3CH_2-OH	ethanol
Ether	R-O-R	$CH_3CH_2-O-CH_2CH_3$	ethoxyether

Table 3.1. (Continued)

Epoxide	$C\!\!-\!\!C$ over O (triangle)	$CH_2\!\!-\!\!CH_2$ over O (triangle)	ethylene oxide
Aldehyde	R-CHO	$CH_3\!-\!\overset{\displaystyle O}{\overset{\|}{C}}\!-\!H$	ethanal
Ketone	R-CO-R	$CH_3\!-\!\overset{\displaystyle O}{\overset{\|}{C}}\!-\!CH_3$	2-propanone
Carboxylic Acid	$R\text{-}CO_2H$	$CH_3\!-\!\overset{\displaystyle O}{\overset{\|}{C}}\!-\!OH$	ethanoic acid
Acid Chloride	R-CO-Cl	$CH_3\!-\!\overset{\displaystyle O}{\overset{\|}{C}}\!-\!Cl$	acetyl chloride
Acid Anhydride	$(RCO)_2O$	$CH_3\!-\!\overset{O}{\overset{\|}{C}}\!-\!O\!-\!\overset{O}{\overset{\|}{C}}\!-\!CH_3$	acetic anhydride
Ester	$R\text{-}CO_2R$	$CH_3\!-\!\overset{\displaystyle O}{\overset{\|}{C}}\!-\!O\!-\!CH_3$	methyl ethanoate
Amide	$R\text{-}CONH_2$	$CH_3\!-\!\overset{\displaystyle O}{\overset{\|}{C}}\!-\!NH_2$	ethanamide
Amine 1°	$R\text{-}NH_2$	$CH_3CH_2\text{-}NH_2$	ethaneamine
Amine 2°	R-NH-R	$CH_3CH_2\text{-}NH\text{-}CH_2CH_3$	diethaneamine
Amine 3°	$R_3\text{-}N$	$(CH_3CH_2)_3\text{-}N$	triethaneamine
Nitro Compound	$R\text{-}NO_2$	$CH_3\!-\!\overset{\oplus}{N}\!\!\big({\overset{\ominus}{O}}\atop{{=}O}\big)$	nitromethane
Nitrile	$R\text{-}C\!\equiv\!N$	$CH_3C\!\equiv\!N$	ethanenitrile
Thiol	R-SH	$CH_3CH_2\text{-}SH$	ethanethiol

3.2   Alkanes

3.2.1   Preparation of Alkanes

3.2.1.1   Wurtz Reaction

$$2\,R\!-\!X \xrightarrow{\ \ Na\ \ } R\!-\!R$$

3.2.1.2   Grignard Reduction

$$RX \ + \ Mg \longrightarrow RMgX \xrightarrow{\ \ H_2O\ \ } RH$$

3.2.1.3   Reduction

$$RX \ + \ Zn \ + \ H^{+} \longrightarrow RH \ + \ ZnX_2$$

$$RX \ + \ LiAlH_4 \xrightarrow{\ \ dry\ ether\ \ } RH \ + \ LiX \ + \ AlX_3$$

3.2.1.4   Kolbe Reaction

$$R\!-\!COO^{\ominus} \xrightarrow[-\,e^{\ominus}]{} R\!-\!R$$

3.2.1.5   Hydrogenation

$$R\!-\!C\!\equiv\!C\!-\!R' \xrightarrow{\ \ H_2\ \ } RCH_2CH_2R'$$

3.2.2   Reactions of Alkanes

3.2.2.1   Combustion

$$R \ + \ O_2 \longrightarrow CO_2 \ + \ H_2O$$

3.2.2.2 Halogenation

$$R + X_2 \xrightarrow[\text{light}]{\text{heat or}} RX + HCl$$

Reactivity   X: $Cl_2 > Br_2$
H: $3° > 2° > 1° > CH_3\text{-}H$

3.2.2.3 Free Radical Substitution

$$X_2 \xrightarrow{\text{heat or light}} 2\ X\cdot$$

$$R\text{—}H + X\cdot \longrightarrow R\cdot + HX$$

$$R\cdot + X_2 \longrightarrow R\text{—}X + X\cdot$$

3.3   Alkenes

3.3.1   Preparation of Alkenes

3.3.1.1   Dehydrohalogenation of Alkyl Halides

Ease of dehydrohalogenation $3° > 2° > 1° >$

3.3.1.2   Dehalogenation of Vicinal Dihalides

3.3.1.3   Dehydration of Alcohols

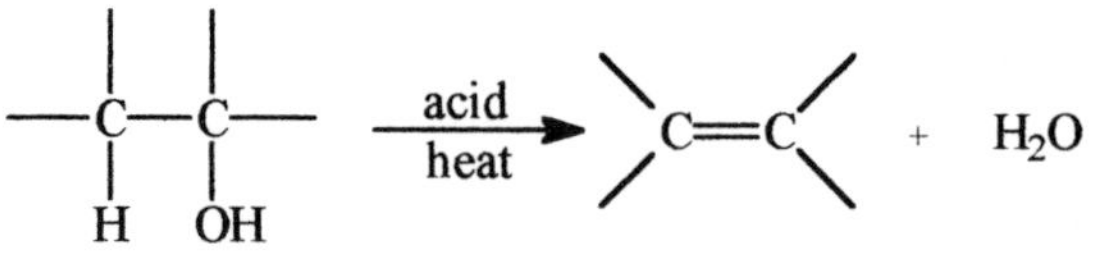

3.3.1.4   Reduction of Alkynes

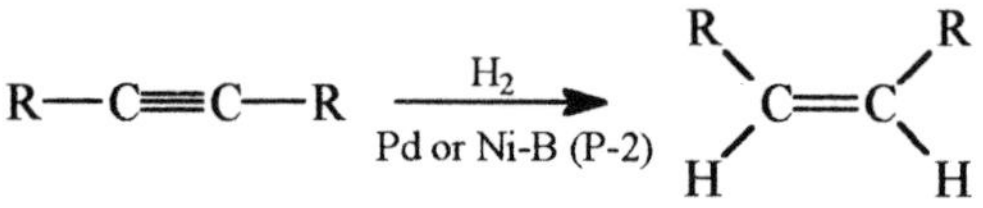

cis

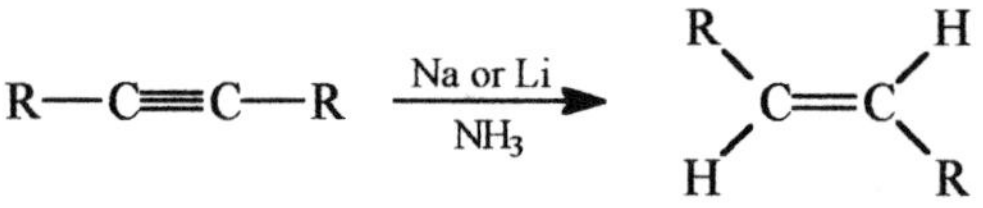

trans

3.3.2   Reactions of Alkenes

3.3.2.1   Hydrogenation

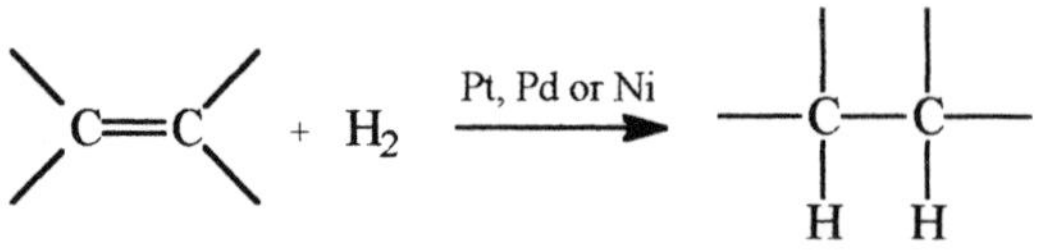

3.3.2.2   Halogenation

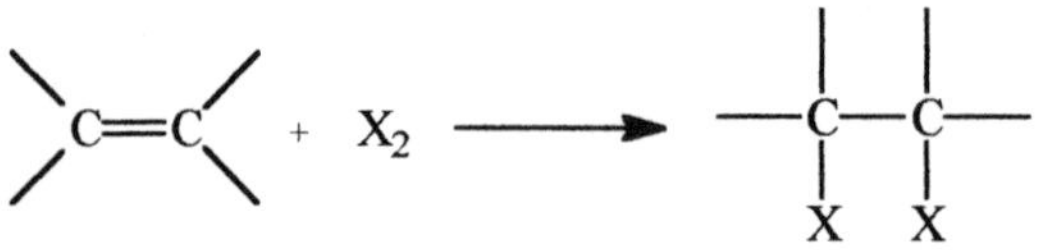

$X_2 = Cl_2, Br_2$

3.3.2.3 Addition of Hydrogen Halide

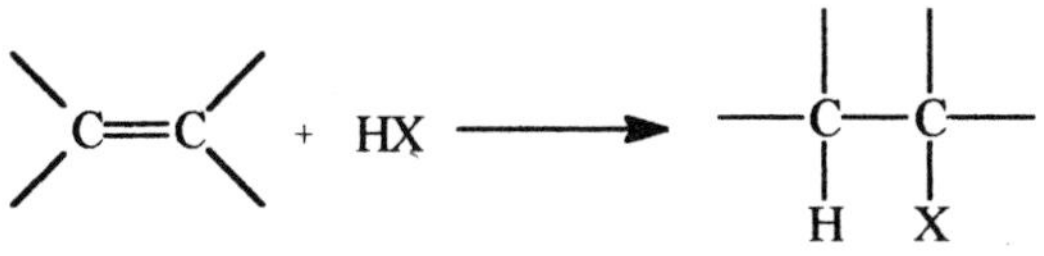

$HX = HCl, HBr, HI$

Markovnikov's rule: The hydrogen of the acid attaches itself to the carbon atom which already has the greatest number of hydrogens. In the presence of peroxide, HBr will undergo anti-Markovnikov addition.

3.3.2.4 Addition of Sulfuric Acid

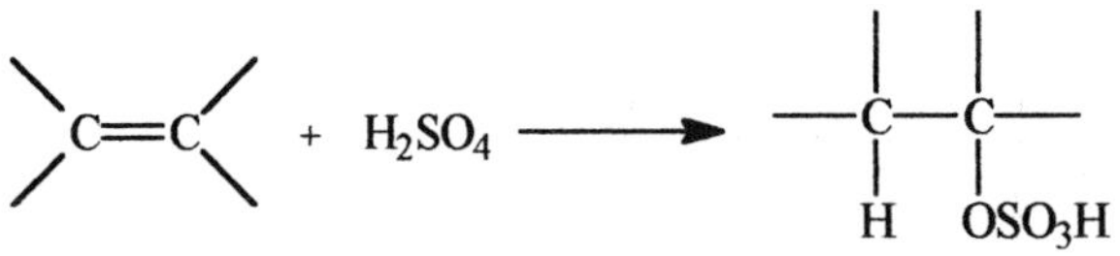

3.3.2.5 Addition of Water

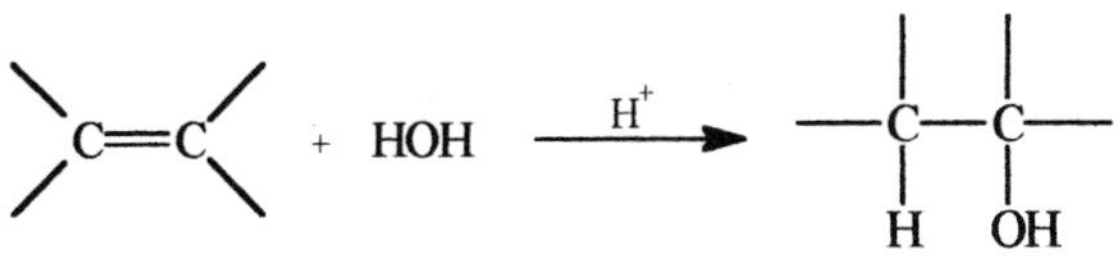

3.3.2.6 Halohydrin Formation

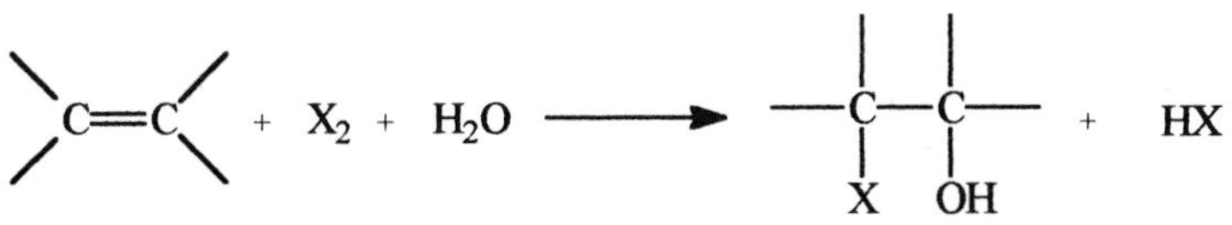

$X_2 = Cl_2, Br_2$

3.3.2.7   Oxymercuration-Demercuration

3.3.2.8   Hydroboration-Oxidation

3.3.2.9   Polymerization

3.3.2.10   Hydroxylation

3.3.2.11   Halogenation - Allylic Substitution

$X_2 = Cl_2, Br_2$

3.4   Dienes

Isolated dienes can be prepared following the methods for alkanes using difunctional starting materials.

3.4.1 Preparation of Conjugated Dienes

3.4.1.1 Dehydration of 1,3-Diol

$$CH_3-\underset{\underset{OH}{|}}{CH}-CH_2-\underset{\underset{OH}{|}}{CH_2} \xrightarrow[\text{acid}]{\text{heat}} CH_2=CH-CH=CH_2$$

3.4.1.2 Dehydrogenation

$$CH_3-CH_2-CH_2-CH_3 \xrightarrow[\text{catalyst}]{\text{heat}} \begin{array}{c} CH_3-CH_2=CH-CH_2 \\ + \\ CH_3-CH=CH-CH_3 \end{array}$$

$$\downarrow \text{heat} \mid \text{catalyst}$$

$$CH_2=CH-CH=CH_2$$

3.4.2 Reactions of Dienes

3.4.2.1 1,4-Addition

$$CH_2=CH-CH=CH_2 + X_2 \longrightarrow \underset{\underset{X}{|}}{CH_2}-CH=CH-\underset{\underset{X}{|}}{CH_2}$$

$X_2 = Cl_2, Br_2$

3.4.2.2 Polymerization

$$n \ \ CH_2=\underset{\underset{CH_3}{|}}{C}-CH=CH_2 \xrightarrow{\text{catalyst}} \left[-CH_2-\underset{\underset{CH_3}{|}}{C}=CH-CH_2- \right]_n$$

3.5 Alkynes

3.5.1 Preparation of Alkynes

3.5.1.1 Dehydrohalogenation of Alkyl Dihalides

3.5.1.2 Dehalogenation of Tetrahalides

3.5.1.3 Reaction of Water and Calcium Carbide

$$CaC_2 + H_2O \longrightarrow CH\equiv CH + Ca(OH)_2$$

3.5.2 Reactions of Alkynes

3.5.2.1 Hydrogenation

$$ —C{\equiv}C— \xrightarrow[\text{NH}_3]{\text{Na or Li}} \quad \begin{array}{c} \diagdown \\ \diagup H \end{array} C{=}C \begin{array}{c} H \\ \diagdown \end{array} $$

3.5.2.2 Halogenation

$$ —C{\equiv}C— \; + \; X_2 \longrightarrow \quad \begin{array}{c} \diagdown \\ X \end{array} C{=}C \begin{array}{c} \diagup \\ X \end{array} \; + \; X_2 \longrightarrow \quad —\underset{X}{\overset{X}{C}}—\underset{X}{\overset{X}{C}}— $$

$X_2 = Cl_2, Br_2$

3.5.2.3 Addition of Hydrogen Halide

$$ —C{\equiv}C— \; + \; HX \longrightarrow \quad \begin{array}{c} \diagdown \\ H \end{array} C{=}C \begin{array}{c} \diagup \\ X \end{array} \; + \; HX \longrightarrow \quad —\underset{H}{\overset{H}{C}}—\underset{X}{\overset{X}{C}}— $$

$X = Cl, Br, I$

3.5.2.4 Addition of Water (Hydration)

$$ —C{\equiv}C— \; + \; H_2O \xrightarrow[\text{HgSO}_4]{\text{H}_2\text{SO}_4} \quad \begin{array}{c} \diagdown \\ H \end{array} C{=}C \begin{array}{c} \diagup \\ OH \end{array} \longleftarrow \quad —\underset{H}{\overset{}{C}}—\underset{O}{\overset{}{C}}— $$

3.6 Benzene

3.6.1 Preparation of Benzene

3.6.1.1 Ring Formation

$$ 3\;H—C{\equiv}C—H \xrightarrow{580^{\circ}C} \quad \bigcirc $$

3.6.1.2 Cyclization

$$CH_3-CH_2-CH_2-CH_2-CH_2-CH_3 \xrightarrow{Cr_2O_3}$$

3.6.1.3 Elimination

—OH $\xrightarrow{\text{Zn dust}}$ + ZnO

3.6.2 Reactions of Benzenes

3.6.2.1 Nitration

+ $HONO_2$ $\xrightarrow{H_2SO_4}$ —NO_2 + H_2O

3.6.2.2 Sulfonation

+ $HOSO_3H$ $\xrightarrow{SO_3}$ —SO_3H + H_2O

3.6.2.3 Halogenation

+ X_2 $\xrightarrow{Fe}$ —X + HX

X = Cl, Br

3.6.2.4 Friedel-Crafts Alkylation

+ RCl $\longrightarrow$ —R + HCl

3.6.2.5   Friedel-Crafts Acylation

$+ ROCl \longrightarrow$ —COR $+ HCl$

3.6.2.6   Hydrogenation

$+ ROCl \longrightarrow$ —COR $+ HCl$

3.6.2.7   Bromination

$+ 3\ Br_2 \xrightarrow{\text{sunlight}}$

3.6.2.8   Combustion

2 $+ (15\text{-}n)O_2 \longrightarrow (12\text{-}n)CO_2 + 6\ H_2O + nC$

3.7   Alkylbenzenes

3.7.1   Preparation of Alkylbenzenes

3.7.1.1   Freidel-Crafts Alkylation

$+$ $RX \xrightarrow{\text{lewis acid}}$ R $+ HX$

Lewis acid: $AlCl_3$, BF_3, HF

Ar-H cannot be used in place of R-X

3.7.1.2   Side Chain Conversion

3.7.1.3   Electrophilic Aromatic Substitution

3.7.1.4   Hydrogenation

3.7.2   Reactions of Alkylbenzenes

3.7.2.1   Hydrogenation

3.7.2.2   Oxidation

3.7.2.3 Substitution in ring - electrophilic aromatic substitution

3.7.2.4 Substitution in the side chain

3.8   Alkenylbenzenes

3.8.1   Preparation of Alkenylbenzenes

3.8.1.1   Dehydrogenation

3.8.1.2   Dehydrohalogenation

3.8.1.3   Dehydration

3.8.2   Reactions of Alkenylbenzenes

3.8.2.1   Cataylic Hydrogenation

3.8.2.2 Oxidation

3.8.2.3 Ring Halogenation

3.9 Alkyl Halides

3.9.1 Preparation of Alkyl Halides

3.9.1.1 From Alcohols

$$R-OH + HX \xrightarrow{H_2SO_4} R-X + H_2O$$

3.9.1.2 Addition of Hydrogen Halide to Alkenes

3.9.1.3 Halogenation of Alkanes

$$R-H + X_2 \longrightarrow R-X + HX$$

3.9.1.4   Halide Exchange

$$R\text{---}X \ + \ NaI \xrightarrow{\ \text{acetone}\ } R\text{---}I \ + \ NaX$$

3.9.1.5   Halogenation of Alkenes and Alkynes

$$X_2 = Cl_2, Br_2$$

$$X_2 = Cl_2, Br_2$$

3.9.2   Reactions of Alkyl Halides

3.9.2.1   Addition of Hydroxide

$$R\text{---}X \ + \ OH^- \longrightarrow R\text{---}OH$$

3.9.2.2   Addition of Water

$$R\text{---}X \ + \ H_2O \longrightarrow R\text{---}OH$$

3.9.2.3   Addition of Alkoxide

$$R\text{---}X \ + \ OR^- \longrightarrow R\text{---}OR$$

3.9.2.4 Addition of Carboxylate

$$R-X + {}^-OOCR' \longrightarrow R-OOCR'$$

3.9.2.5 Addition of Hydrosulfide

$$R-X + SH^- \longrightarrow R-SH$$

3.9.2.6 Addition of Thioalkoxide

$$R-X + SR'^- \longrightarrow R-SR'$$

3.9.2.7 Addition of Sulfide

$$R-X + SR'_2 \longrightarrow R-SR'_2{}^+ X^-$$

3.9.2.8 Addition of Thiocyanide

$$R-X + SCN^- \longrightarrow R-SCN$$

3.9.2.9 Addition of Iodide

$$R-X + I^- \longrightarrow R-I$$

3.9.2.10 Addition of Amide

$$R-X + NH_2^- \longrightarrow R-NH_2$$

3.9.2.11 Addition of Ammonia

$$R-X + NH_3 \longrightarrow R-NH_2$$

3.9.2.12   Addition of 1° Amine

$$R\!-\!X \; + NH_2R' \longrightarrow R\!-\!NHR'$$

3.9.2.13   Addition of 2° Amine

$$R\!-\!X \; + NHR'_2 \longrightarrow R\!-\!NR'_2$$

3.9.2.14   Addition of 3° Amine

$$R\!-\!X \; + \; NR'_2 \longrightarrow R\!-\!NR'^{+}_3 \; X^{-}$$

3.9.2.15   Addition of Azide

$$R\!-\!X \; + \; N_3^{-} \longrightarrow R\!-\!N_3$$

3.9.2.16   Addition of Nitrite

$$R\!-\!X \; + \; NO_2^{-} \longrightarrow R\!-\!NO_2$$

3.9.2.17   Addition of Phosphine

$$R\!-\!X \; + \; P(C_6H_5)_3 \longrightarrow R\!-\!^{+}P(C_6H_5)_3X^{-}$$

3.9.2.18   Addition of Cyanide

$$R\!-\!X \; + \; C\!\equiv\!N^{-} \longrightarrow R\!-\!CN$$

3.9.2.19   Addition of Alkynyl Anion

$$R\!-\!X \; + \; ^{-}C\!\equiv\!C\!-\!R' \longrightarrow R\!-\!C\!\equiv\!C\!-\!R'$$

3.9.2.20 Addition of Carbanion

$$R—X + R'^- \longrightarrow R—R'$$

$$R—X + CH(COOR')_2^- \longrightarrow RCH(COOR')_2$$

$$R—X + CH(COCH_3)(COOR)^- \longrightarrow R—CH(COCH_3)(COOR)$$

$$R—X + Ar—H \xrightarrow{AlCl_3} R—Ar$$

3.10   Aryl Halides

3.10.1   Preparation of Aryl Halides

3.10.1.1   Halogenation by Substitution

$X_2 = Cl_2, Br_2$

3.10.1.2   From Arylthallium Compounds

$$ArH + Tl(OOCCF_3)_3 \longrightarrow ArTl(OOCCF_3)_2 \xrightarrow{KI} ArI$$

3.10.1.3   From Diazonium Salt

$$Ar \xrightarrow[\text{H}_2\text{SO}_4]{\text{HNO}_3} ArNO_2 \xrightarrow{\text{redn}} ArNH_2 \xrightarrow[0^o]{\text{HONO}} ArN_2{}^+$$

$$BF_4^- \quad CuCl \quad CuBr \quad I^-$$

$$ArF \quad ArCl \quad ArBr \quad ArI \quad + \quad N_2$$

3.10.1.4   Halogenation by Addition

$$+ \quad PCl_5 \longrightarrow \qquad + \quad POCl_3$$

3.10.2   Reactions of Aryl Halides

3.10.2.1   Grignard Reagent Formation

$$ArBr + Mg \xrightarrow{\text{dry ether}} ArMgBr$$

$$ArCl + Mg \xrightarrow{\text{THF}} ArMgCl$$

3.10.2.2   Nucleophilic Aromatic Substitution

$$+ \quad Z^- \longrightarrow \qquad + \quad X$$

Z = strong base

3.10.2.3 Electrophilic Aromatic Substitution

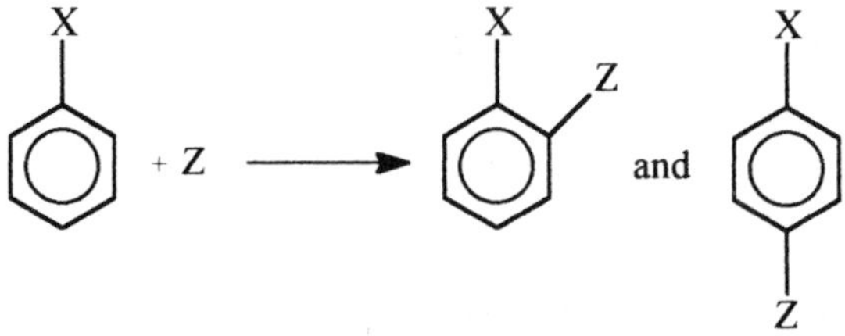

X deactivates and directs ortho, para in electrophilic aromatic substitution.

3.11   Alcohols

3.11.1   Preparation of Alcohols

3.11.1.1 Addition of Hydroxide

$$R—X + NaOH \longrightarrow R—OH + NaX$$

3.11.1.2 Grignard Synthesis

$$H—CHO + R—Mg—X \longrightarrow R—CH_2—O—Mg—X$$

$$R—CH_2—O—Mg—X + HX \longrightarrow R—CH_2—OH + MgX_2$$

primary alcohol

$$R—CHO + R'—Mg—X \longrightarrow R—CHOH—R' + MgX_2$$

secondary alcohol

$$R_2C{=}O + R'—Mg—X + 2\,HX \longrightarrow R_2R'C—OH + MgX_2$$

tertiary alcohol

3.11.1.3 Reduction of Carbonyl Compounds

$$R{-}CHO + Zn + 2\,H_2O \longrightarrow R{-}CH_2{-}OH + Zn^{++} + 2\,OH^-$$

primary alcohol

$$R_2C{=}O + Zn + 2\,H_2O \longrightarrow R_2CHOH + Zn^{++} + 2\,OH^-$$

secondary alcohol

3.11.1.4 Hydration of Alkenes

$$R'{-}CH_2{-}CH_2{-}CH_2{-}R \xrightarrow[\text{cracking}]{} R'{-}CH{=}CH_2 + CH_3R$$

$R > R'$

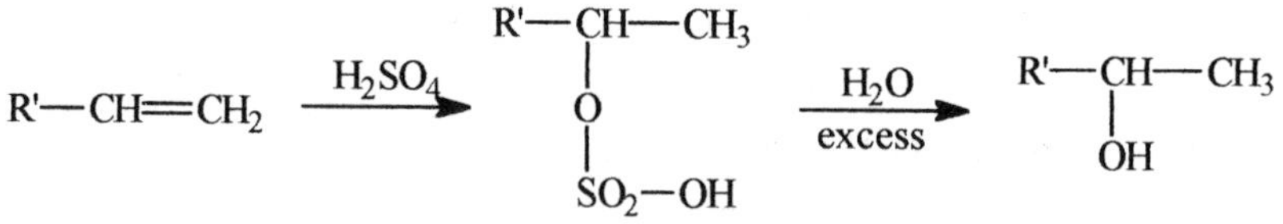

3.11.1.5 Reacton of Amines with Nitrous Acid

$$R{-}NH_2 + HO{-}NO \xrightarrow[H^+]{NaNO_2} ROH + N_2 + H_2O$$

3.11.1.6 Oxymercuration-Demercuration

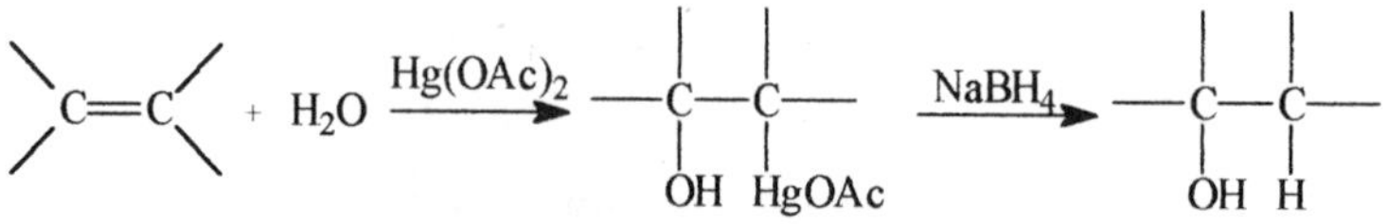

Markovnikov addition

3.11.1.7 Hydroboration-Oxidation

$$\text{C=C} \xrightarrow{\text{(BH}_3)_2} \quad -\overset{|}{\underset{H}{C}}-\overset{|}{\underset{B}{C}}- \quad \xrightarrow[\text{OH}^-]{\text{H}_2\text{O}_2} \quad -\overset{|}{\underset{H}{C}}-\overset{|}{\underset{OH}{C}}-$$

Anti-Markovnikov addition

3.11.2 Reactions of Alcohols

3.11.2.1 Reaction with Hydrogen Halides

$$\text{R—OH} + \text{HX} \longrightarrow \text{RX} + \text{H}_2\text{O}$$

Reactivity of HX: HI > HBr > HCl
Reactivity of ROH: allyl, benzyl >3° >2° >1°

3.11.2.2 Reaction with Phosphorus Trihalide

$$\text{R—OH} + \text{PX}_3 \longrightarrow \text{RX} + \text{H}_3\text{PO}_3$$

3.11.2.3 Dehydration

$$-\overset{|}{\underset{H}{C}}-\overset{|}{\underset{OH}{C}}- \longrightarrow \text{C=C} + \text{H}_2\text{O}$$

3.11.2.4 Ester Formation

$$\text{R—OH} + \text{R'COX} \longrightarrow \text{ROOCR'} + \text{HX}$$

$$\text{R—OH} + \text{R'COOH} \longrightarrow \text{ROOCR'} + \text{H}_2\text{O}$$

3.11.2.5 Reaction with Active Metals

$$R\!-\!OH + M \longrightarrow RO^-M^+ + \tfrac{1}{2}H_2$$

3.11.2.6 Oxidation

$$R\!-\!CH_2OH \xrightarrow{K_2Cr_2O_7} R\!-\!CHO \xrightarrow{K_2Cr_2O_7} RCOOH$$

$$R_2\!-\!CHOH \xrightarrow{K_2Cr_2O_7} R_2\!-\!CHO$$

3.12   Phenols

3.12.1   Preparation of Phenols

3.12.1.1 Nucleophilic Displacement of Halides

3.12.1.2 Oxidation of Cumene

3.12.1.3 Hydrolysis of Diazonium Salts

3.12.1.4 Oxidation of Arylthallium Compounds

3.12.2 Reactions of Phenols

3.12.2.1 Salt Formation

3.12.2.2 Ether Formation - Williamson Synthesis

3.12.2.3 Ester Formation

3.12.2.4   Ring Substution - Nitration

3.12.2.5   Ring Substitution - Sulfonation

3.12.2.6   Ring Substitution - Nitrosation

3.12.2.7 Ring Substitution - Halogenation

3.12.2.8 Ring Substitution - Friedel-Crafts Alkylation

3.12.2.9 Ring Substitution - Friedel-Crafts Acylation

3.12.2.10 Coupling with Diazonium Salts

OH
⬡ + $C_6H_5N_2Cl$ ⟶ HO—⬡—N=N—⬡ + HCl

3.12.2.11 Carbonation. Kolbe Reaction

ONa
⬡ + CO_2 $\xrightarrow[\text{4 - 7 atm}]{125^0 \text{ C}}$ OH ⬡ COONa

3.12.2.12 Carboxylation. Kolbe Reaction

OH
⬡ + CO_2 $\xrightarrow[\text{pressure}]{140^0 \text{ C NaOH}}$ $\xrightarrow{H^+}$ OH ⬡ COOH

3.12.2.13 Aldehyde Formation. Reimer-Tiemann Reaction

OH
⬡ + $CHCl_3$ $\xrightarrow{\text{aq NaOH}}$ OH ⬡ CHO

3.12.2.14 Carboxylic Acid Formation. Reimer-Tiemann Reaction

OH
⬡ $\xrightarrow[\text{NaOH}]{CCl_4}$ $\xrightarrow{H^+}$ OH ⬡ COOH

3.12.2.15 Reaction with Formaldehyde

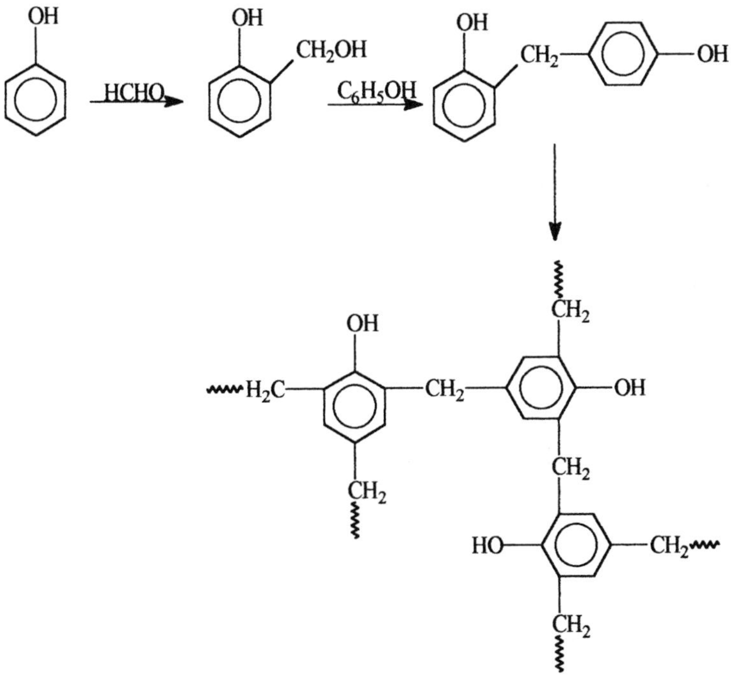

3.13 Ethers

3.13.1 Preparation of Ethers

3.13.1.1 Williamson Synthesis

RX + NaOR' ⟶ ROR' + NaX

R' = alkyl or aryl

3.13.1.2 Oxymercuration-Demercuration

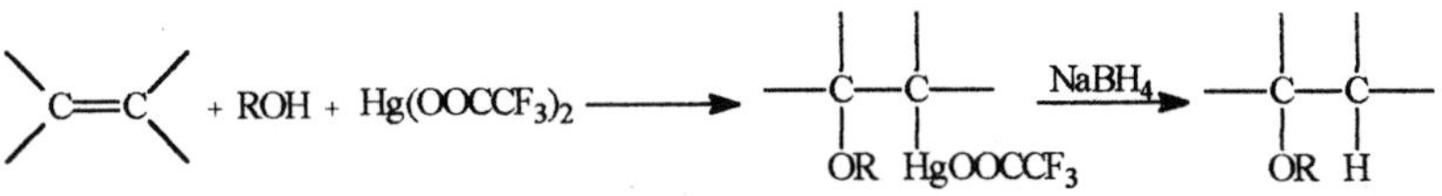

3.13.1.3   Dehydration of Alcohols

$$ROH + HOSO_2OH \longrightarrow ROSO_2OH + H_2O$$

$$2\,ROH \xrightarrow[\text{240-260 C}]{Al_2O_3} ROR + H_2O$$

3.13.2   Reactions of Ethers

3.13.2.1   Single Cleavage by Acids

$$\begin{matrix} ROR' \\ \text{or} \\ ArOR \end{matrix} + HX \longrightarrow \begin{matrix} R'OH \\ \text{or} \\ ArOH \end{matrix} + RX$$

$$ROR + HOSO_2OH \xrightarrow{\text{heat}} ROH + ROSO_2OH$$

$$ROR + HOH \xrightarrow[\text{pressure}]{\text{steam}} 2\,ROH$$

3.13.2.2   Double Cleavage by Acids

$$ROR + PCl_5 \xrightarrow{\text{heat}} 2\,RCl + POCl_3$$

$$ROR + 2\,HI \xrightarrow{\text{heat}} 2\,RI + H_2O$$

$$ROR + 2\,HOSO_2OH \xrightarrow{\text{heat}} 2\,ROSO_2OH + H_2O$$

3.13.2.3 Substitution on the Hydrocarbon Chain

$$R-O-\underset{\underset{H}{|}}{\overset{\overset{H}{|}}{C}}-R \ + \ X_2 \ \longrightarrow \ R-O-\underset{\underset{H}{|}}{\overset{\overset{X}{|}}{C}}-R \ + \ HX$$

HX = Cl, Br

3.14 Epoxides

3.14.1 Preparation of Epoxides

3.14.1.1 Halohydrin Reaction

$$\overset{\diagdown}{\diagup}C=C\overset{\diagup}{\diagdown} \ \xrightarrow[H_2O]{X_2} \ -\underset{\underset{H}{|}}{C}-\underset{\underset{OH}{|}}{C}- \ + \ OH^- \ \longrightarrow \ -\underset{\diagdown O \diagup}{C}-\underset{}{C}-$$

3.14.1.2 Peroxidation

$$\overset{\diagdown}{\diagup}C=C\overset{\diagup}{\diagdown} \ + \ C_6H_5CO_2OH \ \longrightarrow \ -\underset{\diagdown O \diagup}{C}-\underset{}{C}- \ + \ C_6H_5COOH$$

3.14.2 Reactions of Epoxides

3.14.2.1 Acid-Catalyzed Cleavage

$$-\underset{\diagdown O \diagup}{C}-\underset{}{C}- \ + \ H^+ \ \longrightarrow \ -\underset{\underset{Z:}{}}{\overset{\overset{H^+}{\overset{O}{\diagdown\diagup}}}{C}}-\underset{}{C}- \ \longrightarrow \ -\underset{\underset{Z}{|}}{C}-\underset{}{\overset{\overset{OH}{|}}{C}}-$$

3.14.2.2 Base-Catalyzed Cleavage

3.14.2.3 Grignard Reaction

3.15   Aldehydes And Ketones

3.15.1   Preparation of Aldehydes

3.15.1.1 Oxidation

$$RCH_2OH \xrightarrow{K_2Cr_2O_7} R-\overset{\overset{\textstyle H}{|}}{C}=O$$

$$ArCH_3 \begin{cases} \xrightarrow{Cl_2,\ heat} ArCHCl_2 \\ \xrightarrow[ac\ anh]{CrO_3} ArCH(OOCCH_3)_2 \end{cases} \longrightarrow ArCHO$$

3.15.1.2 Reduction

$$RCOCl\ or\ ArCOCl \xrightarrow{LiAlH(OBu\text{-}t)} RCHO\ or\ ArCHO$$

3.15.1.3 Reduction

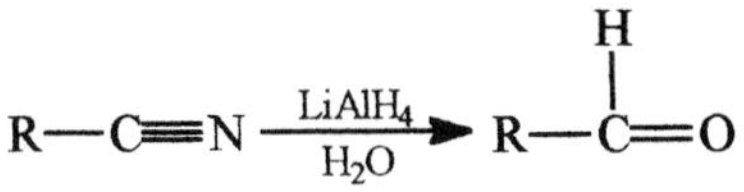

$$R—C≡N \xrightarrow[\text{H}_2\text{O}]{\text{LiAlH}_4} R—\overset{\overset{\displaystyle H}{|}}{C}=O$$

3.15.2 Reactions Specific to Aldehydes

3.15.2.1 Oxidation

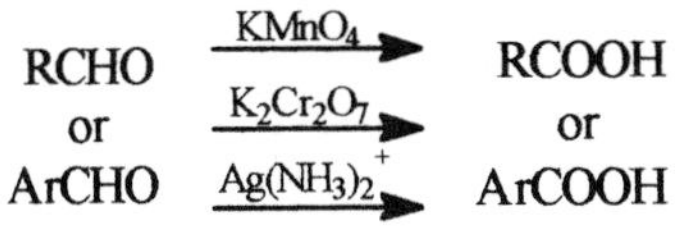

$$
\begin{array}{c}
\text{RCHO} \\
\text{or} \\
\text{ArCHO}
\end{array}
\xrightarrow[\substack{\text{K}_2\text{Cr}_2\text{O}_7 \\ \text{Ag(NH}_3)_2^+}]{\text{KMnO}_4}
\begin{array}{c}
\text{RCOOH} \\
\text{or} \\
\text{ArCOOH}
\end{array}
$$

3.15.2.2 Cannizzaro Reaction

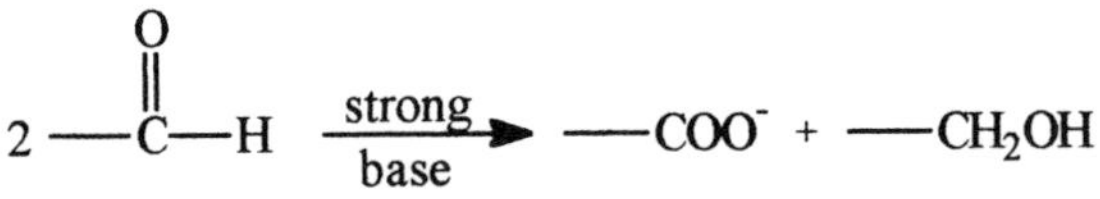

$$2 \; —\overset{\overset{\displaystyle O}{\|}}{C}—H \xrightarrow[\text{base}]{\text{strong}} —COO^- + —CH_2OH$$

3.15.3 Preparation of Ketones

3.15.3.1 Oxidation

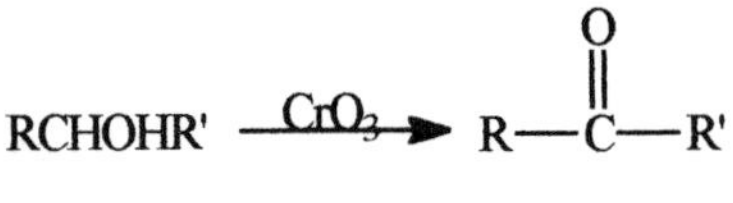

$$RCHOHR' \xrightarrow{\text{CrO}_3} R—\overset{\overset{\displaystyle O}{\|}}{C}—R'$$

3.15.3.2 Friedel-Crafts Acylation

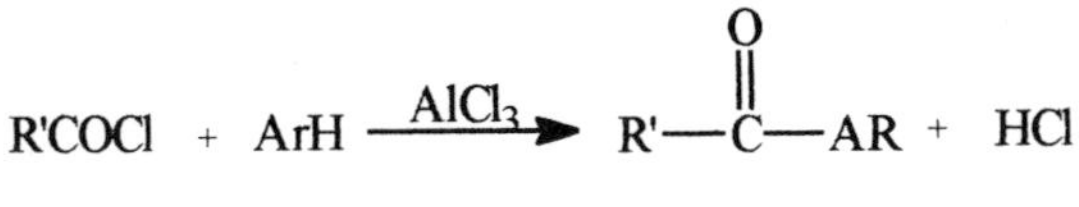

$$R'COCl + ArH \xrightarrow{\text{AlCl}_3} R'—\overset{\overset{\displaystyle O}{\|}}{C}—AR + HCl$$

R' = aryl or alkyl

3.15.3.3 Grignard Reaction

$$R—C\equiv N \xrightarrow{\text{R'MgX}} R—\overset{\overset{\displaystyle O}{\|}}{C}—R'$$

3.15.4   Reactions Specific to Ketones

3.15.4.1 Halogenation

$$\overset{\overset{\displaystyle O}{\|}}{C}—\overset{\overset{\displaystyle |}{\underset{\underset{\displaystyle H}{|}}{C}}}— \;+\; X_2 \longrightarrow \overset{\overset{\displaystyle O}{\|}}{C}—\overset{\overset{\displaystyle |}{\underset{\underset{\displaystyle X}{|}}{C}}}— \;+\; HX$$

3.15.4.2 Oxidation

$$R'—\overset{\overset{\displaystyle O}{\|}}{C}—CH_3 \xrightarrow{\;ox^-\;} R'COO^- \;+\; CHX$$

3.15.5   Reactions Common to Aldehydes And Ketones

3.15.5.1 Reduction to Alcohol

$$\overset{\diagdown}{\underset{\diagup}{}}C{=}O \;-\; \boxed{\begin{array}{c} H_2 \quad Ni,\,Pt,\,Pd \\ \hline LiAlH_4 \\ \text{then } H^+ \end{array}} \;\rightarrow\; —\overset{\overset{\displaystyle |}{\underset{\underset{\displaystyle H}{|}}{C}}}—OH$$

3.15.5.2 Reduction to Hydrocarbon

$$\overset{\diagdown}{\underset{\diagup}{}}C{=}O \xrightarrow{\;Zn(Hg)\;} —\overset{\overset{\displaystyle |}{\underset{\underset{\displaystyle H}{|}}{C}}}—H$$

3.15.5.3 Grignard Reaction

$$\begin{array}{c}\diagdown \\ \diagup\end{array}C{=}O + RMgX \longrightarrow \underset{OMgX}{\overset{C{-}R}{|}} \xrightarrow{H_2O} \underset{OH}{\overset{C{-}R}{|}}$$

3.15.5.4 Cyanohydrin Formation

$$\begin{array}{c}\diagdown \\ \diagup\end{array}C{=}O + CN^- \xrightarrow{H^+} \underset{OH}{\overset{C{-}CN}{|}}$$

3.15.5.5 Addition of Bisulfite

$$\begin{array}{c}\diagdown \\ \diagup\end{array}C{=}O + Na^+HSO_3^- \longrightarrow \underset{OH}{\overset{|}{\underset{|}{C}}}{-}SO_3^-Na^+$$

3.15.5.6 Addition of Ammonia Derivatives

$$\begin{array}{c}\diagdown \\ \diagup\end{array}C{=}O + H_2N{-}G \longrightarrow \underset{OH}{\overset{C{-}NH{-}G}{|}} \longrightarrow \begin{array}{c}\diagdown \\ \diagup\end{array}C{=}N{-}G + H_2O$$

G	Product
R	=NR
OH	=NOH
NH$_2$	=NNH$_2$
NHC$_6$H$_5$	=NNHC$_6$H$_5$
NHCONH$_2$	=NNHCONH$_2$

3.15.5.7 Aldol Condensation

$$\begin{array}{c}\diagdown \\ \diagup\end{array}C{=}O + R'CHO \xrightarrow{OH^-} {-}\underset{}{\overset{OH}{\underset{|}{\overset{|}{C}}}}{-}R'{-}CHO$$

3.15.5.8 Wittig Reaction

$$\diagup\!\!\!\!\diagdown C{=}O \; + \; Ph_3P{=}CRR' \; \longrightarrow \; \diagup\!\!\!\!\diagdown C{=}CRR'$$

3.15.5.9 Acetal Formation

$$\diagup\!\!\!\!\diagdown C{=}O \; + \; 2\,ROH \; \rightleftharpoons \; \overset{\mid}{\underset{\underset{OR}{\mid}}{-C-}}OR \; + \; H_2O$$

3.16   Carboxylic Acids

3.16.1   Preparation of Carboxylic Acids

3.16.1.1 Oxidation of Primary Alcohols

$$RCH_2OH \xrightarrow{\;KMnO_4\;} RCOOH$$

3.16.1.2 Oxidation of Alkylbenzenes

$$Ar{-}R \; \xrightarrow[\;K_2Cr_2O_7\;]{\;KMnO_4\;} \; Ar{-}COOH$$

3.16.1.3 Carbonation of Grignard Reagents

$$\underset{(or\,ArX)}{RX} \xrightarrow{\;Mg\;} RMgX \xrightarrow{\;CO_2\;} RCOMgX \xrightarrow{\;H^+\;} \underset{(or\,ArCOOH)}{RCOOH}$$

3.16.1.4 Hydrolysis of Nitriles

$$\begin{matrix} R{-}C{\equiv}N \\ or \\ Ar{-}C{\equiv}N \end{matrix} + H_2O \xrightarrow[\text{base}]{\text{acid or}} \begin{matrix} R{-}COOH \\ or \\ Ar{-}COOH \end{matrix}$$

3.16.2   Reactions of Carboxylic Acids

3.16.2.1   Acidity Salt Formation

$$RCOOH \longrightarrow RCOO^- + H^+$$

$$ArCOOH \longrightarrow ArCOO^- + H^+$$

3.16.2.2   Conversion to Acid Chloride

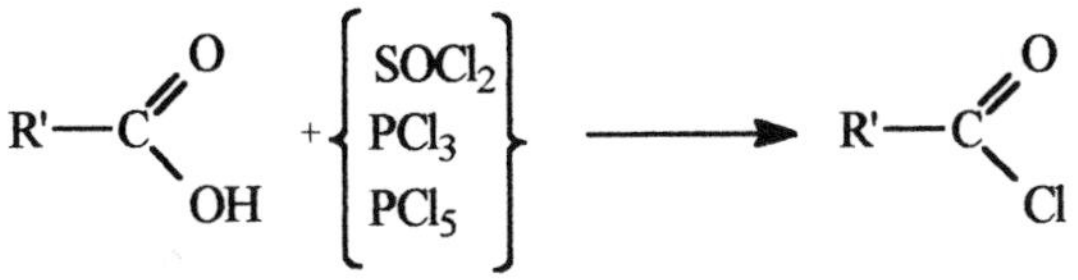

R' = alkyl or aryl

3.16.2.3   Conversion to Esters

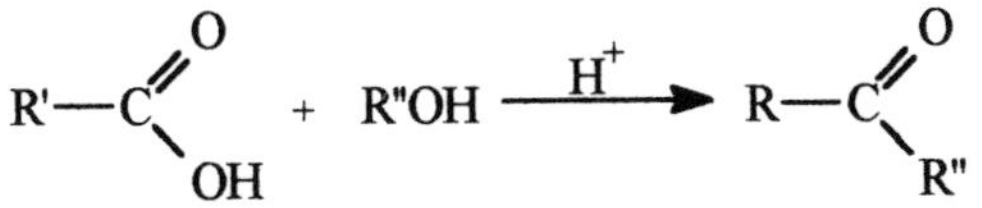

R = alkyl or aryl

3.16.2.4   Conversion to Amides

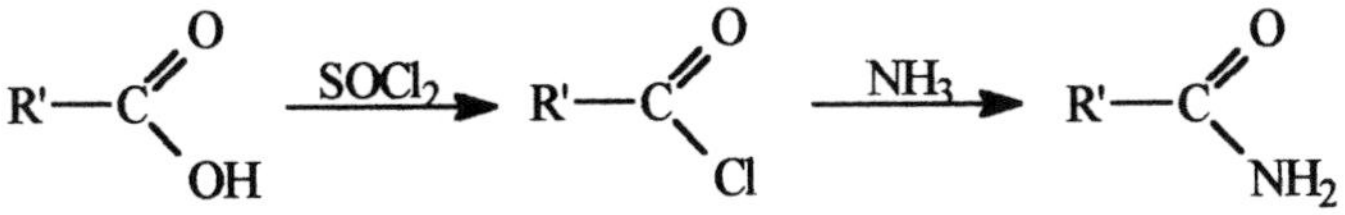

R' = alkyl or aryl

3.16.2.5 Reduction

$$R'COOH \xrightarrow{\text{LiAlH}_4} R'CH_2OH$$

R' = alkyl or aryl

3.16.2.6 Alpha-Halogenation of Aliphatic Acids

$$RCH_2COOH + X_2 \longrightarrow \underset{\underset{X}{|}}{R}CHCOOH + HX$$

3.16.2.7 Ring Substitution in Aromatic Acids

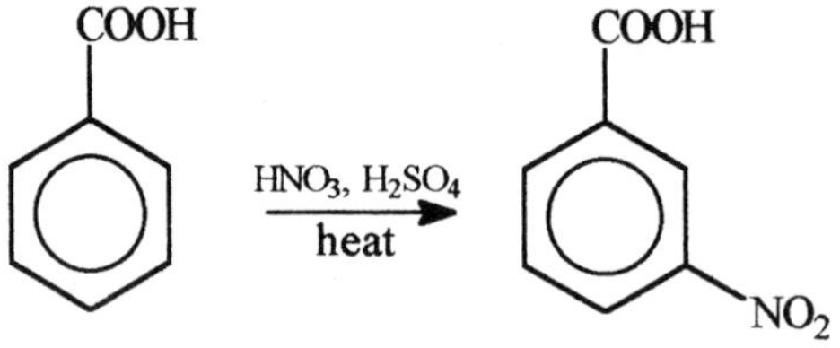

3.17   Acid Chlorides

3.17.1   Preparation of Acid Chlorides

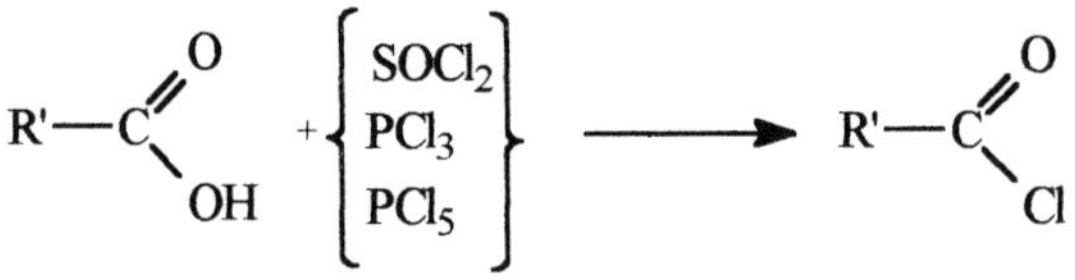

R' = alkyl or aryl

3.17.2   Reactions of Acid Chlorides

3.17.2.1 Hydroysis (Acid Formation)

R'—C(=O)Cl + H_2O ⟶ R'—C(=O)OH + HCl

R' = alkyl or aryl

3.17.2.2 Ammonolysis (Amide Formation)

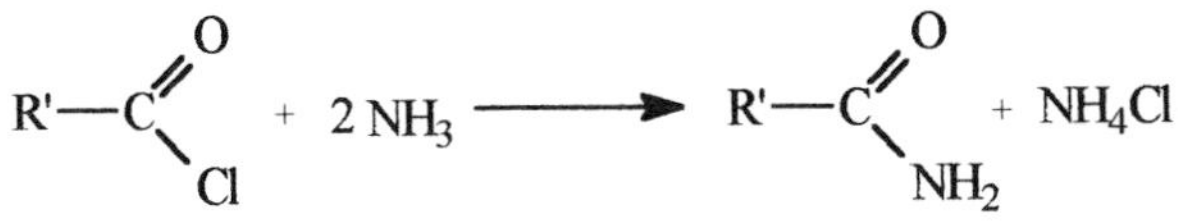

R'—C(=O)Cl + 2 NH_3 ⟶ R'—C(=O)NH_2 + NH_4Cl

R' = alkyl or aryl

3.17.2.3 Alcoholysis (Ester Formation)

R'—C(=O)Cl + R"OH ⟶ R'—C(=O)OR" + HCl

R' = alkyl or aryl

3.17.2.4 Friedel-Crafts Acylation (Ketone Formation)

R'COCl + ArH $\xrightarrow{AlCl_3}$ R'COAr + HCl

R' = alkyl or aryl

3.17.2.5 Ketone Formation by Reaction with Organocadmium Compounds

R'MgX $\xrightarrow{\text{CdCl}_2}$ R'$_2$Cd / RCOCl or ArCOCl $\longrightarrow$ RCOR' or ArCOR'

R' must be an acyl or primary alkyl alcohol

3.17.2.6 Aldehyde Formation by Reduction

RCOCl or ArCOCl $\xrightarrow{\text{LiAlH(OBu-t)}_3}$ RCHO or ArCHO

3.17.2.7 Rosenmund Reduction

$$R-\overset{\overset{\displaystyle O}{\|}}{C}-Cl + H_2 \xrightarrow{Pd/BaSO_4} RCHO + HCl$$

3.17.2.8 Reduction to Alcohols

$$2\,CH_3COCl + LiAlH_4 \longrightarrow LiAlCl_2(COCH_2CH_3)_2 \xrightarrow{H^+} 2\,CH_3CH_2OH$$

3.18   Acid Anhydrides

3.18.1   Preparation of Acid Anhydrides

3.18.1.1 Ketene Reaction

$$CH_2{=}C{=}O + CH_3COOH \longrightarrow CH_3\overset{\overset{\displaystyle O}{\|}}{C}-O-\overset{\overset{\displaystyle O}{\|}}{C}CH_3$$

3.18.1.2 Dehydration of Dicarboxylic Acids

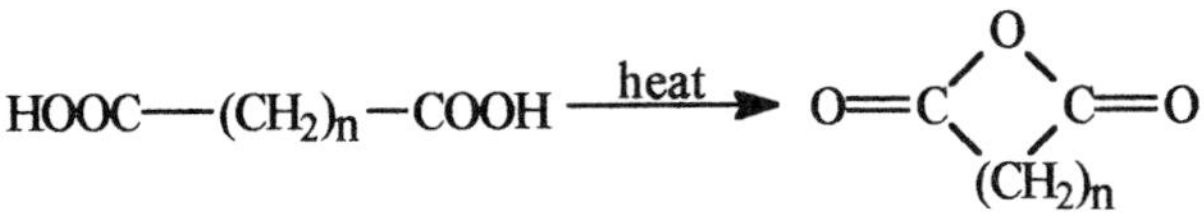

$$HOOC-(CH_2)_n-COOH \xrightarrow{heat} O=C\underset{(CH_2)_n}{\overset{O}{\diagdown\diagup}}C=O$$

$n = 2, 3, 4$

3.18.2 Reactions of Acid Anhydrides

3.18.2.1 Hydroylsis (Acid Formation)

$$\underset{R'C}{\overset{O}{\parallel}}-O-\underset{CR'}{\overset{O}{\parallel}} + H_2O \longrightarrow 2\,R'COOH$$

R' = alkyl or aryl

3.18.2.2 Ammonolysis (Amide Formation)

$$(R'CO)_2O + 2\,NH_3 \longrightarrow R'CONH_2 + R'COO^-NH_4^+$$

R' = alkyl or aryl

3.18.2.3 Alcoholysis (Ester Formation)

$$\underset{R'}{\overset{O}{\parallel}}{-}\underset{C}{\overset{}{}}-O-\underset{C}{\overset{O}{\parallel}}-R' + R''O-H \longrightarrow R'-\underset{C}{\overset{O}{\parallel}}-OR'' + R'COOH$$

R' = alkyl or aryl

3.18.2.4 Friedel-Crafts Acylation (Ketone Formation)

$$(RCO)_2O + ArH \xrightarrow[\text{acid}]{\text{Lewis}} RCOAr + RCOOH$$

3.19   Esters

3.19.1   Preparation of Esters

3.19.1.1   From Acids

$$R'COOH + R''OH \longrightarrow R'COOR'' + H_2O$$

R' = alkyl or aryl

3.19.1.2   From Acid Chlorides

R' = alkyl or aryl

3.19.1.3   From Acid Anhydrides

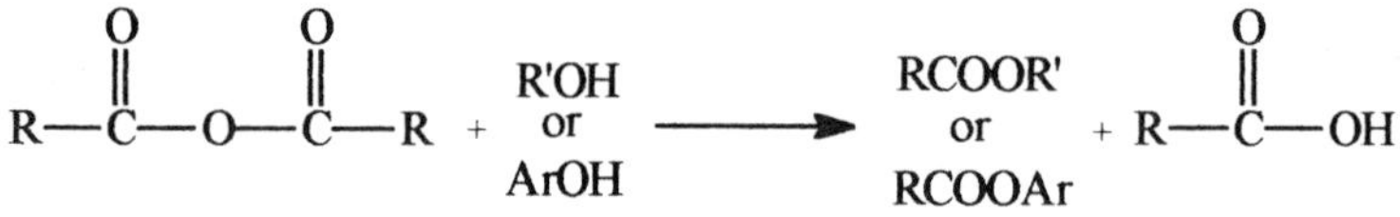

3.19.1.4   Transesterification

3.19.1.5   From Ketene and Alcohols

$$CH_2{=}C{=}O + ROH \longrightarrow CH_3COOR$$

3.19.2   Reactions of Esters

3.19.2.1 Hydrolysis

$$\underset{\text{R'}}{\overset{\overset{\displaystyle O}{\|}}{\text{R'}-\text{C}-\text{OR''}}} + H_2O \xrightarrow{\;H^+\;} \underset{}{\overset{\overset{\displaystyle O}{\|}}{\text{R'}-\text{C}-\text{OH}}} + \text{R''OH}$$

R' = alkyl or aryl

3.19.2.2 Saponification

$$\overset{\overset{\displaystyle O}{\|}}{\text{R'}-\text{C}-\text{OR''}} + H_2O \xrightarrow{\;OH^-\;} \overset{\overset{\displaystyle O}{\|}}{\text{R'}-\text{C}-\text{O}^-} + \text{R''OH} \xrightarrow{\;H^+\;} \text{R'COOH}$$

R' = alkyl or aryl

3.19.2.3 Ammonolysis

$$\overset{\overset{\displaystyle O}{\|}}{\text{R}-\text{C}-\text{OR'}} + NH_3 \longrightarrow \overset{\overset{\displaystyle O}{\|}}{\text{R}-\text{C}-\text{NH}_2} + \text{R'OH}$$

3.19.2.4 Transesterification

$$\overset{\overset{\displaystyle O}{\|}}{\text{R}-\text{C}-\text{OR'}} + \text{R''OH} \longrightarrow \overset{\overset{\displaystyle O}{\|}}{\text{R}-\text{C}-\text{OR''}} + \text{R'OH}$$

3.19.2.5 Grignard Reaction

$$\overset{\overset{\displaystyle O}{\|}}{\text{R}-\text{C}-\text{OR'}} + 2\,\text{R''MgX} \longrightarrow \underset{\underset{\displaystyle OH}{|}}{\overset{\overset{\displaystyle R'}{|}}{\text{R}-\text{C}-\text{R''}}}$$

3.19.2.6 Hydrogenolysis

$$R-\overset{\overset{\text{O}}{\|}}{C}-OR' + 2\,H_2 \xrightarrow[250^{\circ},\,3300\ \text{lb/in}^2]{\text{CuO, CuCr}_2\text{O}_4} RCH_2OH + R'OH$$

3.19.2.7 Bouvaeult - Blanc Method

$$R-\overset{\overset{\text{O}}{\|}}{C}-OR' \xrightarrow[\text{Na}]{\text{alcohol}} RCH_2OH + R'OH$$

3.19.2.8 Chemical Reduction

$$4\,R-\overset{\overset{\text{O}}{\|}}{C}-OR' + 2\,LiAlH_4 \longrightarrow \begin{array}{c} LiAl(OCH_2R)_4 \\ + \\ LiAl(OR')_4 \end{array} \longrightarrow \begin{array}{c} RCH_2OH \\ + \\ R'OH \end{array}$$

3.19.2.9 Claisen Condensation

$$-\overset{\overset{\text{O}}{\|}}{C}-OH + -\overset{\overset{\text{O}}{|}}{\underset{H}{C}}-\overset{\overset{\text{O}}{\|}}{C}-OR' \xrightarrow{^-OC_2H_5} -\overset{\overset{\text{O}}{\|}}{C}-\overset{|}{C}-\overset{\overset{\text{O}}{\|}}{C}-OR'$$

3.20   Amides

3.20.1   Preparation of Amides

3.20.1.1   From Acid Chlorides

$$R'COCl + 2\,NH_3 \longrightarrow R'CONH_2 + NH_4Cl$$

$$R'COCl + R''NH_2 \longrightarrow R'CONHR'' + HCl$$

$$R'COCl + 2\,NHR_2" \longrightarrow R'CONR_2" + R"_2{}^+NH_2Cl^-$$

R' = alkyl or aryl

3.20.1.2 From Acid Anhydrides

$$(RCO)_2O + 2\,NH_3 \longrightarrow RCONH_2 + RCOONH_4$$

$$(RCO)_2O + 2\,R'NH_2 \longrightarrow RCONHR' + R'NH_3{}^+RCO_2{}^-$$

$$(RCO)_2O + NHR_2' \longrightarrow RCONR_2' + RCOOH$$

3.20.1.3 From Esters by Ammonolysis

$$RCOOR' + NH_3 \longrightarrow RCONH_2 + R'OH$$

3.20.1.4 From Carboxylic Acids

$$RCOOH + NH_3 \longrightarrow RCOO^-NH_4{}^+ \longrightarrow RCONH_2 + H_2O$$

3.20.1.5 From Nitriles

$$RC\equiv N + H_2O \xrightarrow{\;heat\;} RCONH_2$$

3.20.1.6 From Ketenes and Amines

$$RNH_2 + CH_2{=}C{=}O \longrightarrow CH_3CONHR$$

3.20.2   Reactions of Amides

3.20.2.1 Hydroylsis

$$R'CONH_2 + H_2O \xrightarrow{\ H^+\ } R'COOH + NH_4^+$$

R' = alkyl or aryl

3.20.2.2 Conversion to Imides

3.20.2.3 Reaction with Nitrous Acid

$$RCONH_2 + ONOH \longrightarrow RCOOH + N_2 + H_2O$$

3.20.2.4 Dehydration

$$RCONH_2 + P_2O_5 \longrightarrow RC\equiv N + 2\,HPO_3$$

3.20.2.5 Reduction

$$RCONH_2 \xrightarrow{\ LiAlH_4\ } \xrightarrow{\ H_2O\ } RCH_2-NH_2$$

3.20.2.6 Hoffman Degradation

$$RCONH_2 + NaOBr + 2\,NaOH \longrightarrow RNH_2 + Na_2CO_3 + NaBr + H_2O$$

3.21 Amines

3.21.1 Preparation of Amines

3.21.1.1 Reduction of Nitro Compounds

$$RNO_2 \xrightarrow[\text{H}_2,\text{ cat}]{\text{metal,H}^+} RNH_2$$

3.21.1.2 Reaction of Ammonia with Halides

$$NH_3 \xrightarrow{\text{RX}} RNH_2 \xrightarrow{\text{RX}} R_2NH \xrightarrow{\text{RX}} R_3N$$

3.21.1.3 Reductive Amination

3.21.1.4   Reduction of Nitriles

$$R'C{\equiv}N \xrightarrow[\text{catalyst}]{H_2} R'CH_2NH_2$$

R' = alkyl or aryl

3.21.1.5   Hofmann Degradation

$$RCONH_2 \xrightarrow{OBr^-} RNH_2$$

R' = alkyl or aryl

3.21.2   Reactions of Amines

3.21.2.1   Alkylation

$$RNH_2 \xrightarrow{RX} R_2NH \xrightarrow{RX} R_3N \xrightarrow{RX} R_4N^+X^-$$

$$ArNH_2 \xrightarrow{RX} ArNHR \xrightarrow{RX} ArNR_2 \xrightarrow{RX} ArNR_3^+X^-$$

3.21.2.2   Salt Formation

$$R'NH_3^+X^- \xrightarrow{HX} R'_2NH_2^+X^- \xrightarrow{HX} R'_3NH^+X^-$$

R' = alkyl or aryl

3.21.2.3   Amide Formation

$$RNH_2 \begin{cases} \xrightarrow{R'COCl} R'CONHR \\ \xrightarrow{ArSO_2Cl} ArSO_2NHR \end{cases}$$

$$R_2NH \longrightarrow \begin{array}{l} \xrightarrow{\text{R'COCl}} R'CONR_2 \\ \xrightarrow{\text{ArSO}_2\text{Cl}} ArSO_2NR_2 \end{array}$$

$$R_3N \longrightarrow \begin{array}{l} \xrightarrow{\text{R'COCl}} \text{No Reaction} \\ \xrightarrow{\text{ArSO}_2\text{Cl}} \text{No Reaction} \end{array}$$

3.21.2.4 Reaction of Amines with Nitrous Acid

$$RNH_2 \xrightarrow{\text{HONO}} [R-N\equiv N^+] \xrightarrow{\text{H}_2\text{O}} N_2$$

$$R_2NH \xrightarrow{\text{HONO}} R_2N-N=O$$

$$ArNH_2 \xrightarrow{\text{HONO}} Ar-N\equiv N^+$$

$$ArNHR \xrightarrow{\text{HONO}} Ar-NR-N=O$$

$$Ar-NR_2 \xrightarrow{\text{HONO}} O=N-Ar-NR_2$$

3.22   Alicyclic Compounds

3.22.1   Preparation of Alicyclic Compounds

3.22.1.1   Cyclization

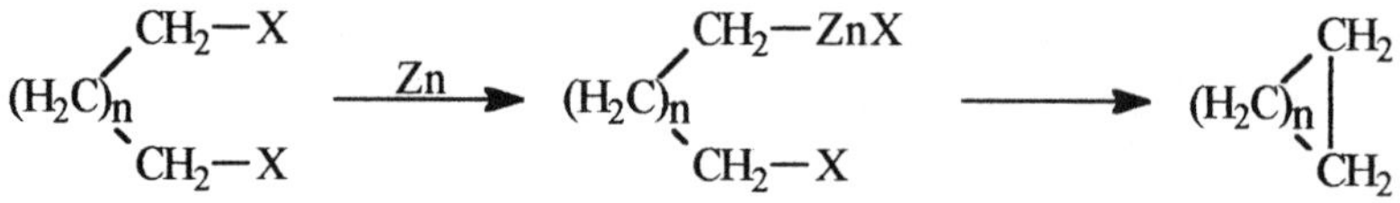

When n = 1 cyclopropane
       n = 2 cyclobutane
       n = 3 cyclopentane

3.22.1.2   Hydrogenation

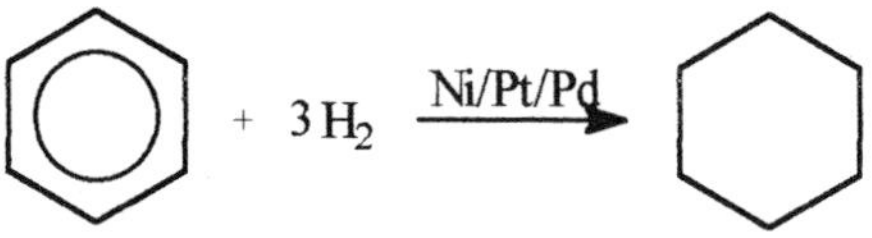

3.22.1.3   Cycloaddition

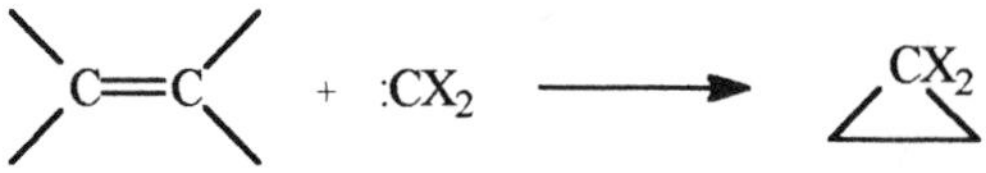

X= H, Cl, Br

3.22.2   Reactions of Alicyclic Compounds

3.22.2.1   Free Radical Addition

3.22.2.2 Addition Reaction

3.23 Heterocyclic Compounds

3.23.1 Preparation of Pyrrole, Furan, and Thiophene

3.23.1.1 Pyrrole

$$HC\equiv CH + 2\ HCHO \xrightarrow{Cu_2C_2} HOCH_2C\equiv CCH_2OH \xrightarrow[\text{pressure}]{NH_3}$$

3.23.1.2 Furan

3.23.1.3 Thiophene

3.23.2   Reactions of Pyrrole, Furan, and Thiophene

3.23.2.1   Pyrrole

3.23.2.2   Furan

3.23.2.3   Thiophene

$$\text{thiophene} + I_2 \xrightarrow[\text{benzene}]{\text{HgO}} \text{2-iodothiophene}$$

$$\text{thiophene} + (CH_3CO)_2O \xrightarrow{H_3PO_4} \text{thiophene-COCH}_3$$

$$\text{thiophene} + CH_3COONO_2 \xrightarrow[-10^{\circ}C]{Ac_2O} \text{thiophene-NO}_2 + \text{3-nitrothiophene (NO}_2)$$

3.23.3 Preparation of Pyridine, Quinoline, and Isoquinoline

3.23.3.1 Pyridine

$$2\ CH_2=CH-CHO + NH_3 \longrightarrow \text{3-methylpyridine (CH}_3)$$

$$\text{butadiene} + N\equiv C-R \longrightarrow \text{dihydropyridine-R} \xrightarrow{K_3[Fe(CN)_6]} \text{pyridine}$$

3.23.3.2 Quinoline

$$\text{aniline (NH}_2) + \begin{matrix} CH_2OH \\ | \\ CHOH \\ | \\ CH_2OH \end{matrix} + C_6H_5NO_2 \xrightarrow[\text{FeSO}_4]{H_2SO_4} \text{quinoline} + C_6H_5NH_2 + H_2O$$

3.23.3.3 Isoquinoine

3.23.4 Reactions of Pyridine, Quinoline, and Isoquinoline

3.23.4.1 Pyridine

3.23.4.2 Quinoline

3.23.4.3 Isoquinoline

3.24   Isomers

3.24.1   Isomers and Stereoisomers

Organic compounds that have the same chemical formula but are attached to one another in different ways are called **isomers**. Isomers that have the same chemical formula and are attached to one another in same way but whose orientation to one another differ are called **Stereoisomers**. There are several different types of isomers that are encountered in organic chemistry.

To represent three dimensional structures on paper the chiral center of a molecule is taken at the cross point of a cross and the groups are attached at the ends. The horizontal line represents the bonds projecting out of the plane of the paper. The vertical line represents the bonds projecting into the plane of the paper.

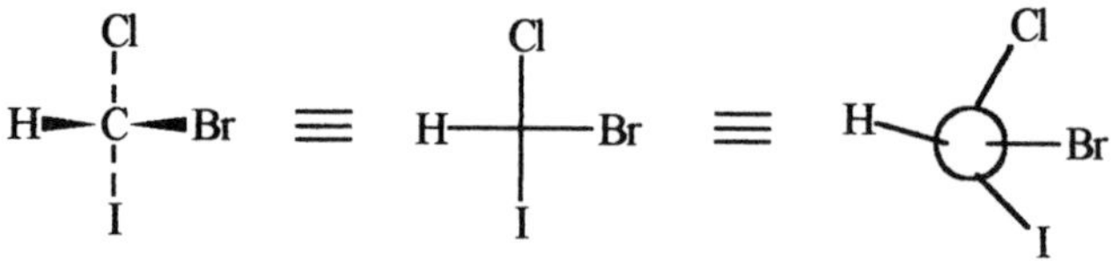

Figure 3.2. Three Dimensional Representations

3.24.2   Optical Activity

In addition to having different arrangements of atoms, certain organic compounds exhibit a unique property of rotating **plane-polarized light** (light that has its amplitude in one plane). Compounds that rotate light are said to be **optically active**. Optically active compounds that rotate light to the right are called dextrorotatory and are symbolized by D or +. Compounds that rotate light to the left are called levorotatory and are symbolized by L or -.

3.24.3   Enantiomers

Enantiomers are stereoisomers that are non-superimposable mirror images of each other. Enantiomers have identical physical and chemical (except towards optically active reagents) properties except for the direction in which plane-polarized light is rotated. Enantiomers account for a compound's optical activity.

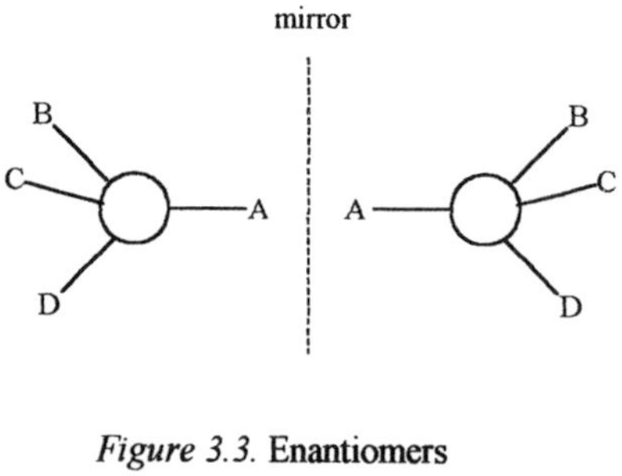

Figure 3.3. Enantiomers

3.24.4   Chirality

Molecules that are not superimposable on their mirror images are **chiral**. Chiral molecules exists as enantiomers but achiral molecules cannot exist as enantiomers. A carbon atom to which four different groups are attached is a **chiral center**. Not all molecules that contain a chiral center are chiral. Not all chiral molecules contain a chiral center.

3.24.5   Racemic Mixture

A mixture of equal parts of enantiomers is called a **racemic mixture**. A racemic mixture contains equal parts of D and L componets and therefore, the mixture is optically inactive.

3.24.6   Diastereomers

Diastereomers are stereoisomers that are not mirror images of each other. Diastereomers have different physical properties. And they maybe dextrorotatory, levorotatory or inactive.

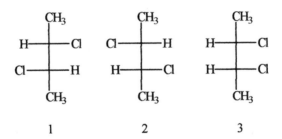

Figure 3.4 Enantiomers and Diastereomers

Structures 1 and 2 are enantiomers, structure 3 is a diastereomer of structures 1 and 2.

3.24.7   Meso Compounds

Meso compounds are superimposable mirror images of each other, even though they contain a chiral center.

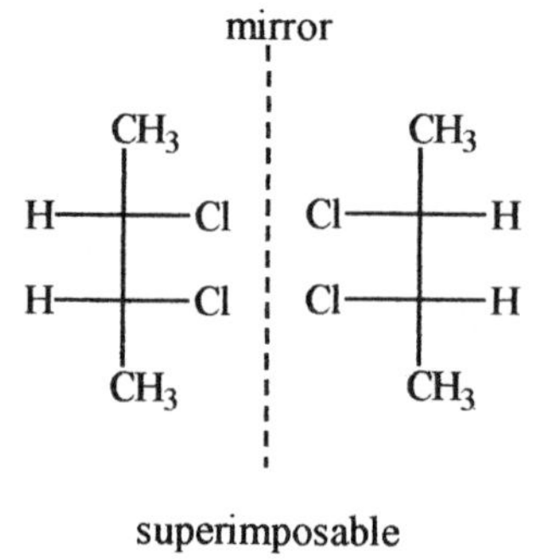

mirror

$$
\begin{array}{cccc}
 & CH_3 & & CH_3 \\
H\!-\!\!\!&\!\!\!-Cl & Cl\!-\!\!\!&\!\!\!-H \\
H\!-\!\!\!&\!\!\!-Cl & Cl\!-\!\!\!&\!\!\!-H \\
 & CH_3 & & CH_3
\end{array}
$$

superimposable

Figure 3.5. Meso Compound

Meso compounds can be recognized by the fact that half the molecule is a mirror image of the other half.

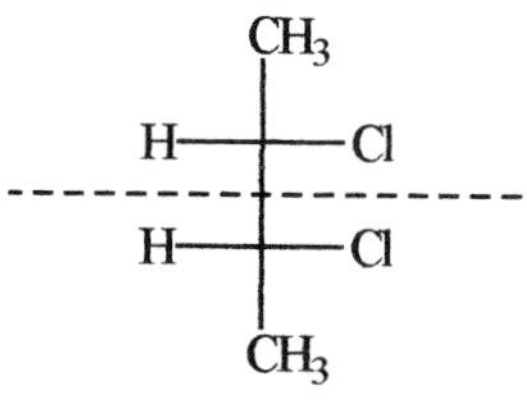

$$
\begin{array}{cc}
 & CH_3 \\
H\!-\!\!\!&\!\!\!-Cl \\
\text{- - - - - - - - - - -} \\
H\!-\!\!\!&\!\!\!-Cl \\
 & CH_3
\end{array}
$$

Figure 3.6. Plane of symmetry of a meso compound

The upper half of the molecule is a non-superimposable mirror image of the lower half, making the top half an enantiomer of the lower half. However, since the two halfs are in the same molecule the rotation of plane-polarized light by the upper half is cancelled by the lower half and the compound is optically inactive.

3.24.8   Positional Isomers

Positional isomers are compounds that have the same number and kind of atoms but are arranged (or bonded) in a different order. They also have different physical and chemical properties. Butane (figure 3.1) can have two different structures, n-butane and 2-methylpropane:

$$
\begin{array}{cc}
 & CH_3 \\
 & | \\
H_3C\!-\!CH_2\!-\!CH_2\!-\!CH_3 & H_3C\!-\!CH\!-\!CH_3
\end{array}
$$

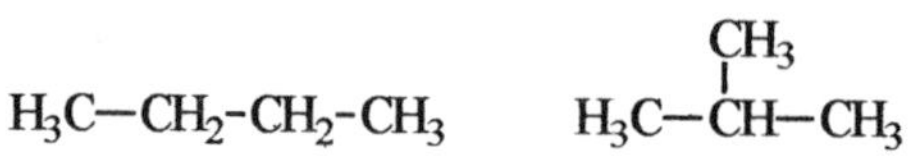

Figure 3.7. n-butane and 2-methylpropane

3.24.9 Geometric Isomers

Geometric isomers or cis-trans isomerism can exist in compounds that contain a double bond or a ring structure. In order for this type of isomerism to exist the groups coming off the same end of the double bond must be different. For example bromoethene does not have cis-trans isomerism.

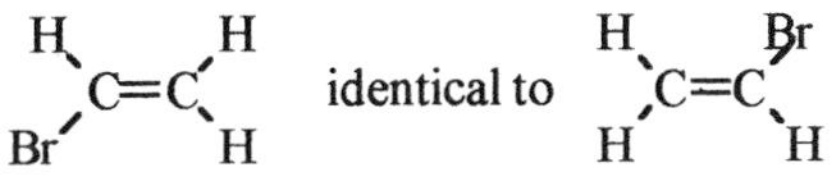

Figure 3.8. Bromoethene structure

However, 1,2-dibromoethene can exists as cis-1,2-dibromoethene and trans-1,2-dibromoethene

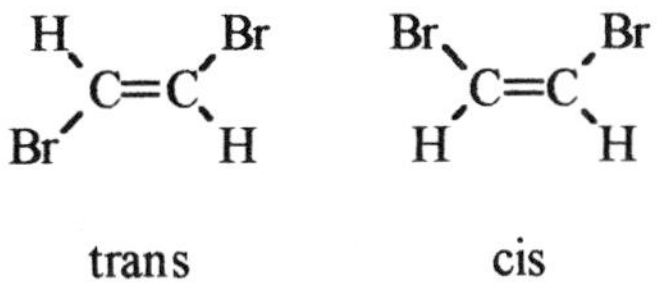

trans   cis

Figure 3.9. Trans and cis butene

Ring structures confer restricted rotation around the bonds and thereby give rise to geometric isomer. In trans-1,3-dichlorocyclopentane one chlorine is above the plane of the ring and on is below. In cis-1,3 dichlorocyclopentane both chlorines are above the plane of the ring.

trans   cis

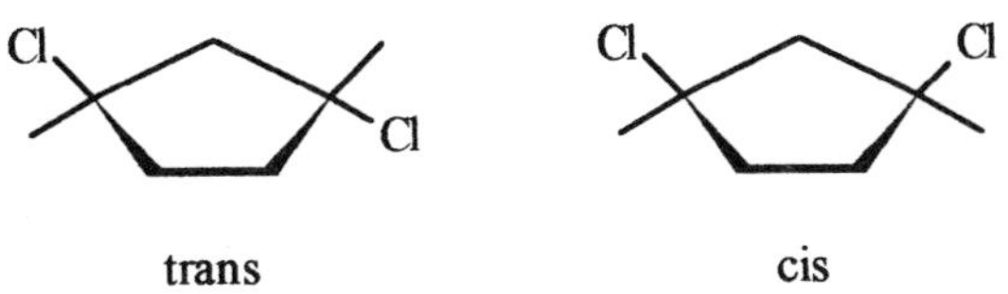

Figure 3.10. Trans and cis-1,3-cyclopropane

3.24.10 Conformational Isomers

Conformational isomers deal with the orientations within a molecule. The free rotation around a single bond accounts for the different conformations that can exist within a molecule. For example, n-butane can have the following conformations:

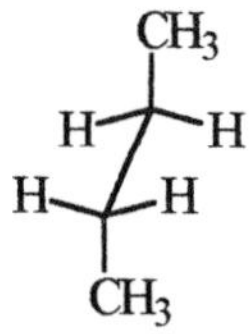

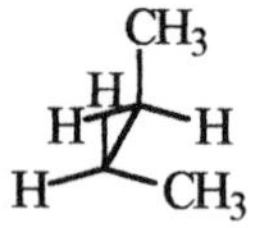

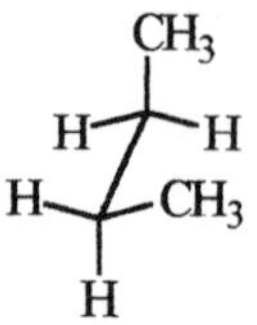

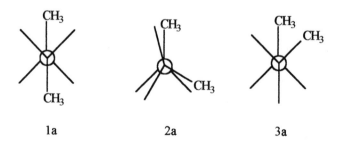

Figure 3.11. Conformational isomers of butane

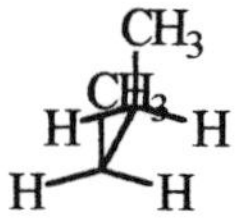

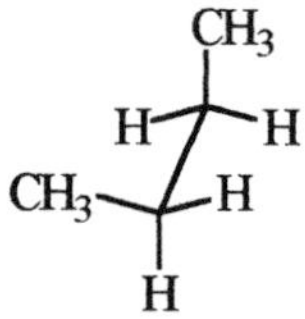

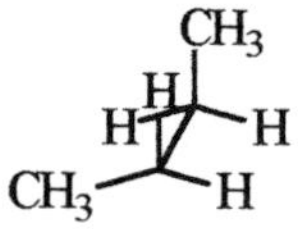

Figure 1 has a *staggered* or *anti* conformation. Since in figure 1, the two methyl groups are farthest apart, this form is referred to as *anti*. Figures 3 and 5 have *staggered* or *gauche* conformations. Figures 2, 4, and 6 have an *eclipsed* conformation.

A different type of projection used to view isomers, called Newman projections, is sometimes used. The following figures are the same as above:

Figure 3.12. Newman projections

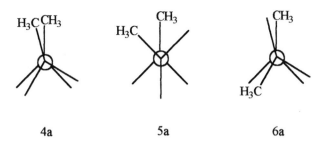

4a   5a   6a

Figure 3.12. Newman projections

Although cyclohexane is a ring structure it does have free rotation around single bonds. Cyclohexane has two main conformations. The most stable form is called the chair form, the les stable is called the boat form:

chair   boat

Figure 3.13. Chair and boat conformations of cyclohexane

The bonds in cyclohexane occupy two kinds of position, six hydrogens lie in the plane and six hydrogens lie either above or below the plane. Those that are in the plane of the ring lie in the "equator" of it, and are called the **equatorial bonds**. Those bonds that are above or below are pointed along the axis perpendicular to the plane and are called **axial bonds**.

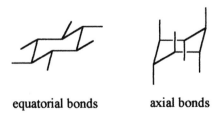

equatorial bonds   axial bonds

Figure 3.14. Equatorial and axial bonds of cyclohexane

3.24.11   Configurational Isomers

The arrangement of atoms that characterizes a certain stereoisomer is called its configuration. In general, optically active compounds can have more than one configuration. Determination of the configuration can be determined by the following two steps:

Step 1. Following a set of sequence rule we assign a sequence of priority to the four atoms attached to the chiral center.
Step 2. The molecule is oriented so that the group of lowest priority is directed away from us. The arrangement of the remaining groups is then observed. If the sequence of highest priority to lowest priority is clockwise, the configuration is designated R. If the sequence is counterclockwise, the configuration is designated S.
From these steps a set of sequence rules can be formulated that will allow a configuration to be designated as either R or S.
Sequence 1. If the four atoms attached to the chiral center are all different, priority depends on the atomic number, with the atom of higher atomic number getting higher priority.
Sequence 2. If the relative priority of two groups cannot be decided by Sequence 1, it shall be determined by a similar comparison of the atoms attached to it.
For example, bromochloroiodomethane, CHClBrI, has two possible configurations as shown in figure 3.8. Using the sequence rules, the order of the atoms for the configuration is I, Br, Cl, H. However, the figure on the left has a different sequence than the one on the right. Hence, (R)-bromochloroiodomethane and (S)-bromochloroiodomethane

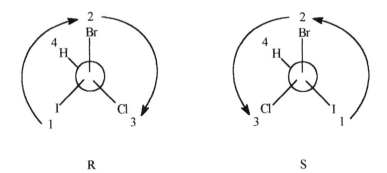

R   S

Figure 3.15. R and S configuration

3.25   Polymer Structures

The following listing of common polymers provides respective structure. The reader should note that the name of the polymer often provides the key to its representative structure. There are, however, names such as polycarbonate that can represent a variety of polymeric materials.

Acrylonitrile-butadiene-styrene terpolymer (ABS):

Buna-N:
    Elastomeric copolymer of butadiene and acrylonitrile.

Buna-S:
    Elastomeric copolymer of butadiene and styrene.

Butyl rubber:

Cellulose:

Epoxy resins:

Ethylene-methacrylic acid copolymers (ionomers):

Ethylene-propylene elastomers:

Formaldehyde resins:
Phenol-formaldehyde (PF):

Urea-formaldehyde (UF):

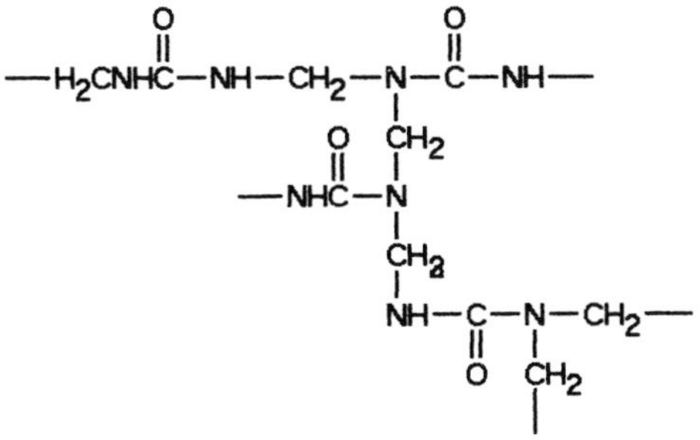

Melamine-formaldehyde (MF):

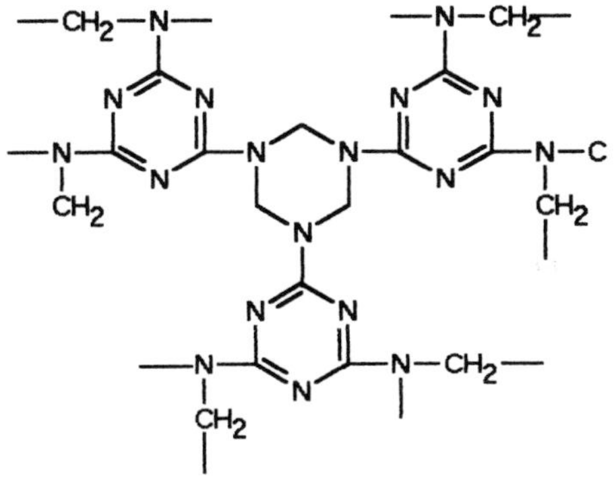

Nitrile rubber (NBR):

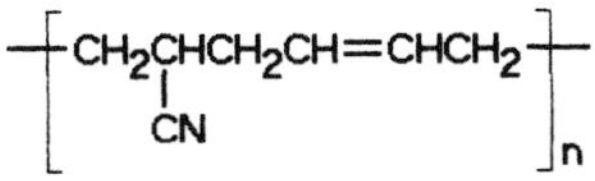

Polyacrylamide:

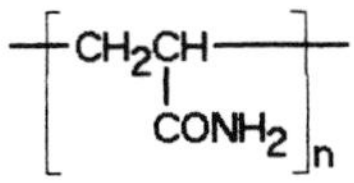

Polyacrylonitrile:

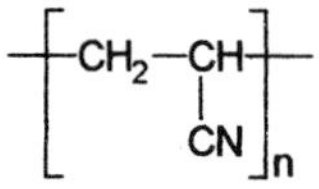

Polyamides (nylons):
Nylon 6,6 and nylon 6:

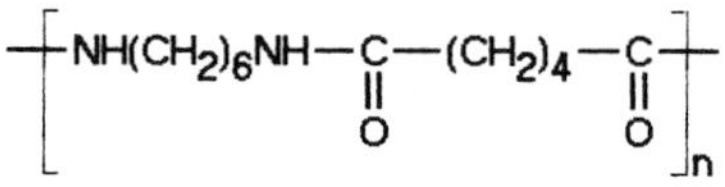

nylon 6,6 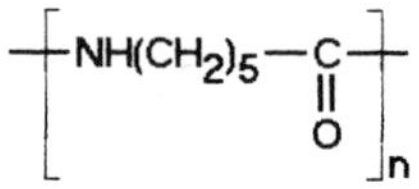   nylon 6

Aromatic polyamides:

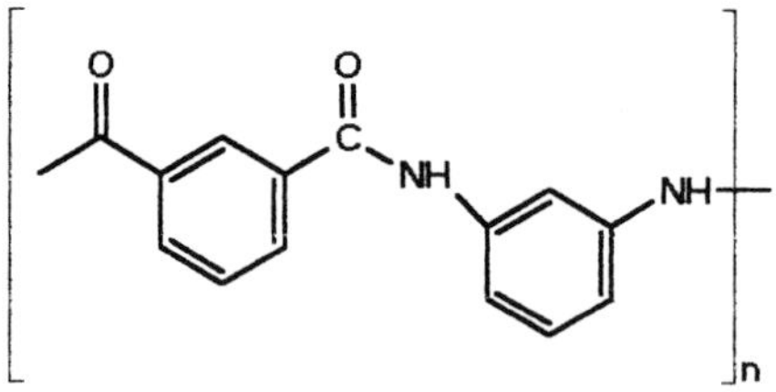

Nomex

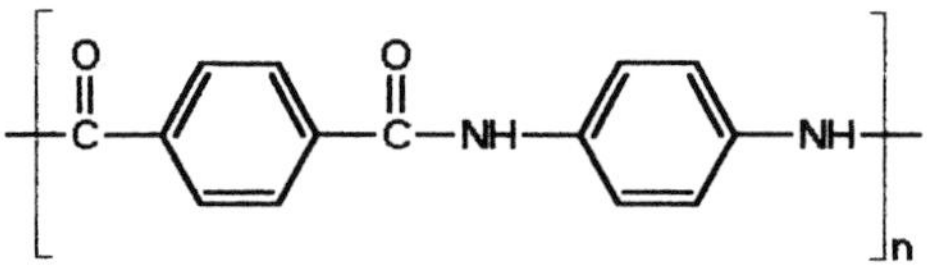

Kevlar

Polyamide imides and polyimides:

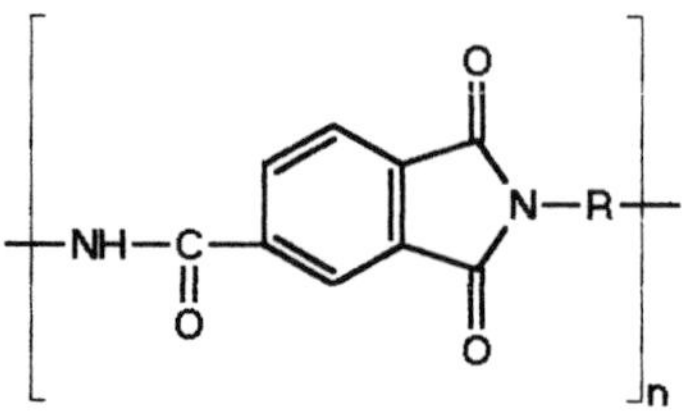

poly(amide imides) 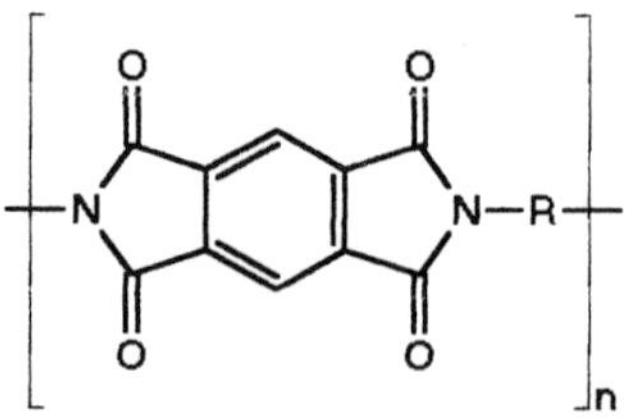   polyimides

Polybutadiene (butadiene rubber, BR):

$$\left[\!-CH_2CH\!=\!CHCH_2-\!\right]_n$$

Polycarbonate (PC):

Polychloroprene:

Polyesters:
Poly(ethylene terephthalate) (PET):

Poly(butylene terephthalate) (PBT):

Aromatic polyesters:

Polyether (polyoxymethylene; polyacetal):

$$\left[OCH_2 \right]_n$$

Polyethylene (PE):

$$\left[CH_2CH_2 \right]_n$$

Low-density polyetylene (LDPE).
High-density polyethylene (HDPE).
Linear low-density polyethylene (LLDPE).
Ultrahigh molecular weight polyethylene (UMWPE).

Poly(ethylene glycol) (PEG):

$$HOCH_2CH_2 \left[OCH_2CH_2 \right]_n OH$$

Polyisobutylene (PIB):

$$\left[CH_2 - \underset{\underset{CH_3}{|}}{\overset{\overset{CH_3}{|}}{C}} \right]_n$$

Polyisoprene:

$$\left[CH_2 - \underset{\underset{CH_3}{|}}{C} = CH - CH_2 \right]_n$$

Poly(methyl methacrylate) (PMMA):

$$\left[CH_2 - \underset{\underset{COOCH_3}{|}}{\overset{\overset{CH_3}{|}}{C}} \right]_n$$

Poly(phenylene oxide) (PPO):

poly(2,6-dimethyl-p-phenylene ether)

Poly(phenylene sulfide) (PPS):

Polyphosphazenes:

Polypropylene (PP):

Polystyrene (PS):

Polysulfone:

Polytetrafluoroethylene (PTFE):

Polyurethane:

Poly(vinyl acetate) (PVA):

Poly(vinyl alcohol) (PVAL):

Poly(vinyl butyral) (PVB):

Poly(vinyl carbazole):

Poly(vinyl chloride) (PVC):

Poly(vinyl formal) (PVF):

Poly(vinylidene chloride):

Poly(vinyl pyridine):

Poly(vinyl pyrrolidone):

Silicones (siloxanes):

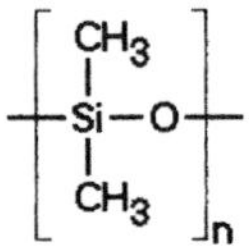

$$\left[\begin{array}{c} CH_3 \\ | \\ -Si-O- \\ | \\ CH_3 \end{array}\right]_n$$

Starch:

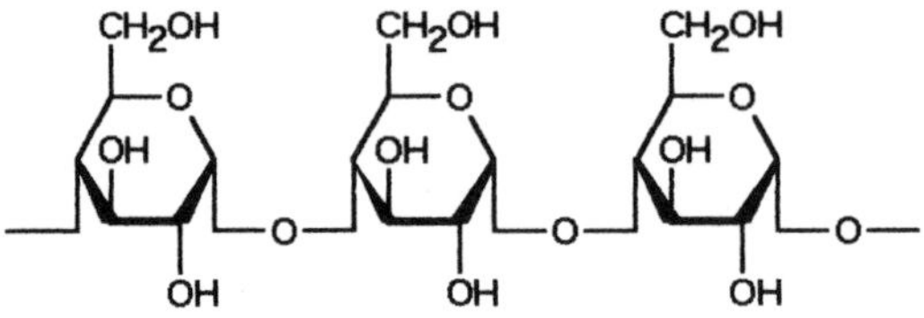

linear amylose

Styrene-acrylonitrile copolymer (SAN):

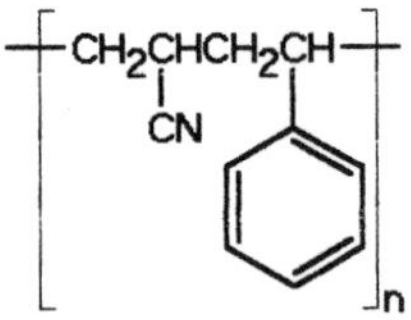

Styrene-butadiene rubber (SBR):

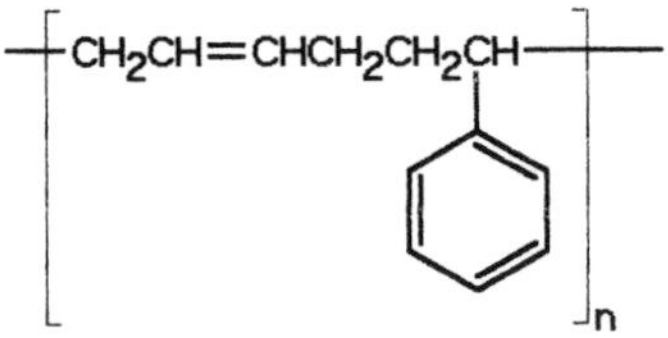

Chapter 4

Instrumental Analysis
4.1 The Electromagnetic Spectrum
4.2 Ultraviolet-Visible Spectrum
4.3 Infrared Spectrum
4.4 Nuclear Magnetic Resonance
4.5 Mass Spectroscopy

4.1   The Electromagnetic Sprectrum

4.1.1   The Electromagnetic Spectrum

The various regions of the electromagnetic spectrum are somewhat arbitrary and not very sharply defined. Each region overlaps at both ends the adjacent regions. The units defining the different regions can be in frequency (Hz, cm^{-1}), energy (kcal/mole, eV) and wavelengths (m and λ, "official" or most common unit used for a given region). Some useful relationships and constants are the following: 1 nanometer (nm) = 1 millimicron (mμ) ,=10^{-9} m; 1 μm (formerly micron) = 10^{-6} m = 10 mm; 1 angstrom (Å) = 0.1 nm = 10^{-8} cm; 1 eV = 23.06 kcal/mole = 8063 cm^{-1}. Energy, E = hv [h = Planck's constant = 9.534 x 10^{-14} kcal-sec/mole = 1.583 x 10^{-34} cal-sec/molecule; v in Hz (cycles/sec)]. E in kcal/mole = 2.8635/λ (λ in μm); E in eV = 12,3451 (λ in Å).

Table 4.1 The Electromagnetic Spectrum

Region	Spectral Use	Hz	cm-1	kcal/mole	eV	m	λ
Cosmic rays	Nuclear transitions	10^{22}	3×10^{11}	9.5×10^{8}	4.1×10^{7}	3×10^{-14}	3×10^{-4} Å
Gamma rays		10^{19}	3×10^{7}	9.5×10^{5}	4.1×10^{4}	3×10^{-10}	3 Å
Soft X-rays	Inner electron transitions	10^{17}	1×10^{5}	9.5×10^{3}	4.1×10^{2}	3×10^{-8}	300 Å
Vacuum uv	Valance electron transitions	1.5×10^{15}	5×10^{4}	143	6.2	2×10^{-7}	200 nm
Quartz uv	(electronic	7.5×10^{14}	2.5×10^{4}	71.5	3.1	4×10^{-7}	400 nm
Visible	spectra)	3.8×10^{14}	1.2×10^{4}	36.2	1.6	8×10^{-7}	800 nm
Near ir	Vibrational ir overtone and combination region	1.2×10^{14}	4×10^{3}	12	0.52	2.5×10^{-6}	2.5 μm
Vibrational ir	Fundamental region	2.5×10^{13}	8×10^{2}	2.4	0.10	1.2×10^{-5}	12.5 μm
Far ir	Sketal, ring torsional, solid state (lattice) deformations	10^{12}	33	9.5×10^{-2}	4.1×10^{-3}	3×10^{-4}	300 μm
Microwave	Molecular rotations; bond torsions	10^{9}	3.3×10^{-2}	9.5×10^{-5}	4.1×10^{-6}	3×10^{-1}	300 mm
Short Radio	Spin orientations	1.5×10^{6}	5×10^{-5}	1.4×10^{-7}	6.1×10^{-9}	2×10^{2}	200 m
Broadcast	Radio, tv, etc.	5.5×10^{5}	1.8×10^{-5}	5.2×10^{-8}	2.2×10^{-9}	5.6×10^{2}	550 m
Long Radio	Induction heating; longwave communication	3×10^{3}	10^{-7}	2.9×10^{-10}	1.3×10^{-11}	10^{5}	10^{5} m
Electric Power	Commercial power and light	3×10^{-1}	10^{-11}	2.9×10^{-14}	1.3×10^{-14}	10^{9}	10^{9} m

4.2 Ultraviolet-Visible Spectroscopy

4.2.1 Solvents

Table 4.2. UV-Vis Solvents

Substance	Formula	BP (°C)	Cut-off (nm) 1 mm	Cut-off (nm) 10 mm
2-Methylbutane	$C_4H_9CH_3$	28	---	210
Pentane	C_5H_{12}	36	---	200
Hexane	C_6H_{14}	69	190	200
Heptane	C_7H_{16}	96	---	195
i-Octane	C_8H_{18}	98	190	205
Cyclopentane	C_5H_{10}	49	---	210
Methylcyclopentane	$C_5H_9CH_3$	72	---	200
Cyclohexane	C_6H_{12}	81	190	195
Methylcyclohexane	$C_6H_{11}CH_3$	101		205
Benzene	C_6H_6	80	275	280
Toluene	$C_6H_5CH_3$	110	280	285
m-Xylene	$C_6H_4(CH_3)_2$	139	285	290
Decalin	$C_{10}H_{18}$	190	---	200
Water	HOH	100	187	191
Methanol	CH_3OH	65	200	205
Ethanol	CH_3CH_2OH	78	200	205
2-Propanol	$CH_3CH(OH)CH_3$	81	200	210
Glycerol	$(HOCH_2)_2CH(OH)$	290d	200	205
Sulfuric acid	96% H_2SO_4	300	---	210
Diethyl ether	Et_2O	35	205	215
THF	$CH_2CH_2CH_2CH_2O$	67	---	220
1,4-Dioxan	$(OCH_2CH_2)_2$	102	210	220
Dibutyl ether	$(C_4H_9)_2O$	142	---	210
Carbon disulfide	CS_2	46	---	380
Chloroform	$CHCl_3$	60	230	245
Carbon tetrachloride	CCl_4	76	245	260
Methylene chloride	CH_2Cl_2	40	220	230
1,1-Dichloroethane	$ClCH_2CH_2Cl$	84	220	230
1,1,2,2-Dichloroethene	$Cl_2C=CCl_2$	120	280	290
Tribromomethane	$CHBr_3$	150	315	330
Bromotrichloromethane	$BrCCl_3$	105	320	340
Acetonitrile	CH_3CN	81	190	195
Methylnitrite	CH_3NO_2	100	360	380
Pyridine	C_5H_5N	112	300	305
N,N-DMF	$HCON(CH_3)_2$	152	---	270
DMSO	$(CH_3)_2SO$	189d	250	265
Acetone	CH_3COCH_3	56	320	330

4.2.2 Woodward's Rules for Diene Absorption

Parent heteroannular diene   214
Parent homoannular diene   253
Add for each substituent:
    Double bond extending conjugation   30
    Alkyl substituent, ring residue, etc.   5
    Exocyclic double bond   5
    $N(alkyl)_2$ 60
    S(alkyl)   30
    O(alkyl)   6
    OAc   0

$$\lambda_{calc\ max} = \text{TOTAL, nm}$$

4.2.3 Selected UV-Vis Tables

Table 4.3. Characteristics of Simple Chromophoric Groups

Chromophore	Example	λ_{max}, mμ	ε_{max}	Solvent
C=C	Ethylene	171	15,530	Vapor
	1-Octene	177	12,600	Heptane
C≡C	2-Octyne	178	10,000	Heptane
		196	2,100	Heptane
		223	160	Heptane
C=O	Acetaldehyde	160	20,000	Vapor
		180	10,000	Vapor
		290	17	Hexane
	Acetone	166	16,000	Vapor
		189	900	Hexane
		279	15	Hexane
$-CO_2H$	Acetic acid	208	32	Ethanol
-COCl	Acetyl chloride	220	100	Hexane
$-CONH_2$	Acetamide	178	9,500	Hexane
		220	63	Water
$-CO_2R$	Ethyl acetate	211	57	Ethanol
$-NO_2$	Nitromethane	201	5,000	Methanol
		274	17	Methanol
$-ONO_2$	Butyl nitrate	270	17	Ethanol
-ONO	Butyl nitrite	220	14,500	Hexane
		356	87	Hexane
-NO	Nitrosobutane	300	100	Ether
		665	20	Ether
C=N	*neo*-Pentylidene n-butylamine	235	100	Ethanol
-C≡N	Acetonitrile	167	weak	Vapor
$-N_3$	Azidoacetic ester	285	20	Ethanol
$=N_2$	Diazomethane	410	3	Vapor
-N=N-	Azomethane	338	4	Ethanol

Table 4.4. Characteristics of Simple Conjugated Chromophoric Groups

Chromophore	Example	λ_{max}, mμ	ε_{max}	Solvent
C=C-C=C	Butadiene	217	20,900	Hexane
C=C-C≡C	Vinylacetylene	219	7,600	Hexane
		228	7,800	Hexane
C=C-C=O	Crotonaldehyde	218	18,000	Ethanol
		320	30	Ethanol
	3-Penten-2-one	224	9,750	Ethanol
		314	38	Ethanol
-C≡C-C=O	1-Hexyn-3-one	214	4,500	Ethanol
		308	20	Ethanol
C=C-CO$_2$H	*cis*-Crotonic acid	206	13,500	Ethanol
		242	250	Ethanol
-C≡C-CO$_2$H	*n*-Butylpropiolic acid	210	6,000	Ethanol
C=C-C=N-	*N-n*-Butylcrotonaldimine	219	25,000	Hexane
C=C=C≡N	Methacrylonitrile	215	680	Ethanol
C=C=NO$_2$	1-Nitro-1-propene	229	9,400	Ethanol
		235	9,800	Ethanol
HO$_2$C-CO$_2$H	Oxalic acid	185	4,000	Water

Table 4.5. Ultraviolet Absorption of Some Monosubstituted Benzenes (in Water)

C$_6$H$_5$X	Primary Band		Secondary Band	
X=	λ_{max}, mμ	ε_{max}	λ_{max}, mμ	ε_{max}
-H	203.5	7,400	254	204
-NH$_3$	203	7,500	254	169
-CH$_3$	206.5	7,000	261	225
-I	207	7,000	257	700
-Cl	209.5	7,400	263.5	190
-Br	210	7,900	261	192
-OH	210.5	6,200	270	1,450
-OCH$_3$	217	6,400	269	1,480
-SO$_2$NH$_2$	217.5	9,700	264.5	740
-CN	224	13,000	271	1,000
-CO$_2^-$	224	8,700	268	560
-CO$_2$H	230	11,600	273	970
-NH$_2$	230	8,600	280	1,430
-O$^-$	235	9,400	287	2,600
-NHCOCH$_3$	238	10,500	---	---
-COCH$_3$	245.5	9,800	---	---
-CHO	249.5	11,400	---	---
-NO$_2$	268.5	7,800	---	---

4.3 Infrared Spectroscopy

4.3.1 Infrared Media

Table 4.6. IR media

	Usable Regions of the Spectrum:cm^{-1} (μm)		
Substance	Near IR[1]	Mid IR	Far IR
KBr	10000 - 3333 (1-3)	5000 - 667 (2-15)	800 - 250 (12.5-40)
KCl	10000 - 3333 (1-3)	5000 - 667 (2-15)	800 - 526 (12.5-19)
CsBr	10000 - 3333 (1-3)	5000 - 667 (2-15)	800 - 250 (12.5-40)
CsI	10000 - 3333 (1-3)	5000 - 667 (2-15)	800 - 130 (12.5-77)
AgCl	10000 - 3333 (1-3)	5000 - 667 (2-15)	800 - 530 (12.5-19)
TlCl	10000 - 3333 (1-3)	5000 - 667 (2-15)	800 - 530 (12.5-19)
Polyethylene	10000 - 3333 (1-3)[2]	2500 - 1540 (4-6.5)	625 - 278 (16-36)
		1250 - 741 (8-13.5)	
Polystyrene			400 - 278 (25-36)[3]
Teflon		5000 - 1333 (2-7.5)[4]	
		1111 - 690 (9-14.5)	
Kel-F		5000 - 1333 (2-7.5)	
		870 - 690 (11.5-14.5)	
Nujol[5]		5000 - 3333 (2-3)	667 - 286 (15-35)
		2500 - 1540 (4-6.5)	
		1250 - 667 (8-15)	
Fluorolube[6]		5000 - 1430 (2-7)	
C_4Cl_6[7]		5000 - 1667 (2-6)	
		1430 - 1250 (7-8)	

(From The Chemist's Companion, Gordon & Ford, Copyright © 1972 John Wiley & Sons, Inc. Reprinted with permission of John Wiley & Sons, Inc.)

1. Very little work has been reported for near ir spectra of pellets or mulls; most measurements are made in solution.
2. For a 0.1 mm sheet when compensated; otherwise, interfering bands occur at 1.9, 2.3, and 2.6 - 2.8 μm.
3. For 0.025 mm sheet.
4. For 0.01 mm thick specimen; Teflon and Kel-F powders are very good for observing details in the OH stretch region.
5. Common brand of heavy mineral oil.
6. A saturated chlorofluorocarbon oil.
7. Hexachloro-1,3-butadiene.

4.3.2   Infrared Absorption Frequencies of Functional Groups

Table 4.7. Saturated Compounds   sp = sharp, br = broad, (w) = weak, (s) = strong

Functional Group	Absorption Range (cm^{-1})	Example (cm^{-1})	Example Compound
SATURATED COMPOUNDS			
Linear			
CH_3 asymmetric	2970-2950	2967	*n*-Octane
CH_3 symmetric	2885-2865	2868	
CH_3 asymmetric	1465-1440	1466	
CH_3 symmetric	1380-1370	1380	
CH_2 asymmetric	2930-2915	2920	*n*-Octane
CH_2 symmetric	2860-2840	2854	
CH_2	1480-1450	1470	
$(CH_2)n, n > 4$	723-720	723	*n*-Octane
	735-725	733	*n*-Pentane
	755-735	741	2-Methylpentane
	800-770	781	*n*-Propane
Branched			
CH	2890	2890	Triphenylmethane
	1340	1341	
$(CH_3)_2$-CH-	1385-1380	1384	2-Methylheptane
	1372-1366	1366	
	1175-1165		
	1160-1140		
	922-917		
$(CH_3)_3$-C-	1395-1380	1393	2,2-Dimethylhexane
	1375-1365	1366	
	1252-1245		
	1225-1195		
	930-925		
-C(CH_3)-C(CH_3)-	1165-1150	1160	3,4-Dimethylhexane
	1130-1120	1122	
	1080-1065	1071	
$(CH_3)_2$-C-R_2	1391-1381	1389	3,3-Dimethylhexane
	1220-1190	1192	
	1195-1185	1189	
$(C_2H_5)_2$-CH-R	1250	1250	3-Ethylhexane
	1150	1155	
	1130	1131	
C-C(CH_3)-C	1160-1150		

Table 4.7. (Continued)

Cyclic Compounds			
Cyclopropane derivatives	3100-3072	3075	Cyclopropane
	3033-2995	3028	
	1030-1000	1024	
Cyclobutane derivatives	3000-2975	2974	Cyclobutane
	2924-2874	2896	
	1000-960 or		
	930-890	901	
Cyclopentane derivatives	2959-2952	2951	Cyclopentane
	2870-2853	2871	
	1000-960	968	
	930-890	894	
Cyclohexane derivatives	1055-1000	1038	Cyclohexane
	1015-950	1014	

Table 4.8. Unsaturated Compounds

Functional Group	Absorption Range (cm^{-1})	Example (cm^{-1})	Example Compound
UNSATURATED COMPOUNDS			
Isolated -C=C- bonds			
CH$_2$=CH-	3095-3075	3096	1-Butene
	3030-2990	2994	
	1648-1638	1645	
	1420-1410	1420	
	1000-980	994	
	915-905	912	
CH$_2$=C	3095-3075	3096	Methylpropene
	1660-1640	1661	
	1420-1410	1420	
	895-885	887	
-CH=C	3040-3010	3037	3-Methyl-2-pentene
	1680-1665	1675	
	1350-1340	1351	
	840-805	812	
-CH=CH- (*cis*)	3040-3010	3030	*cis*-2-Butene
	1660-1640	1661	
	1420-1395	1406	
	730-675	675	

Table 4.8. (Continued)

	3040-3010	3021	*trans*-2-Butene
-CH=CH-(*trans*)	1700-1670	1701	
	1310-1295	1302	
	980-960	964	
Conjugated -C=C- bonds			
-C=C-C=C-	1629-1590	1592	1,3-Butadiene
	1820-1790	1821	
Allenic -C=C- bonds			
-C=C=C-	1960-1940		
	1070-1060		
-C≡C- bonds			
-C≡C-	2270-2250	2268	2-Pentyne
-C≡CH groups			
CH (stretch)	3320-3300*	3320	1-Butyne
-C≡C-	2140-2100	2122	
CH (bend)	700-600		
* CCl₄ Solutions only			

Table 4.9. Aromatic Compounds

Functional Group	Absorption Range (cm^{-1})	Example (cm^{-1})	Example Compound
AROMATIC COMPOUNDS			
General			
CH	3060-3010		
CH substitution bands			
overtones	2000-1650 (w)		
C=C	1620-1590 sp		
	1590-1560 sp		
CH	1510-1480 sp		
	1450 sp		
Mono-substitution			
	1175-1125	1170	Toluene
	1110-1070	1088	
	1070-1000	1032	
	765-725	728 (s)	
	720-690	693 (s)	
Di-substitution			
ortho	1225-1175	1185	*o*-Xylene
	1125-1090	1121	
	1070-1000	1053	
	765-735	741 (s)	
meta	1175-1125	1171	*m*-Xylene
	1110-1070	1095	
	1070-1000	1039	
	900-770	769 (s)	
	710-690	690 (s)	

Table 4.9. (Continued)

para	1225-1175	1219	*p*-Xylene
	1125-1090	1120	
	1070-1000	1043	
	855-790	796 (s)	
Tri-substitution			
1,2,3-	1175-1125	1162	1,2,3-Trimethylbenzene
	1110-1070	1095	
	1000-960	1009	
	800-755	765 (s)	
	740-695	710 (s)	
1,2,4-	1225-1175	1156	1,2,4-Trimethylbenzene
	1130-1090	1130	
	1000-960	1000	
	900-865	873 (s)	
	855-800	805 (s)	
1,3,5-	1175-1125	1165	1,3,5-Trimethylbenzene
	1070-1000	1039	
	860-810	836 (s)	
	705-685	690 (s)	

Table 4.10. Alcohol Compounds

Functional Group	Absorption Range (cm^{-1})	Example (cm^{-1})	Example Compound
ALCOHOLS			
General			
OH unbridged group	3650-3590 sp		
OH inter- and intra-	3570-3450		
molecularly H - bonded			
OH intermolecularly	3400-3200 br		
H-bonded			
Primary alcohols			
	1350-1260	1339	
	1065-1020	1028	1-Pentanol
Secondary alcohols			
	1370-1260	1369	
	1120-1080	1111	2-Pentanol
Tertiary alcohols			
	1410-1310	1379	
	1170-1120	1124	2-Methylbutanol
Aromatic ring hydroxy compounds			
OH unbrided	3617-3599 sp		
OH dimer	3460-3322 br		
OH polymer	3370-3322 br		
	1410-1310	1350	
	1225-1175	1225	Phenol

Table 4.11. Peroxide Compounds

Functional Group	Absorption Range (cm^{-1})	Example (cm^{-1})	Example Compound
PEROXIDES **Aliphatic**			
	1820-1810		
	1800-1780		
	890-820		
Aromatic			
	1805-1780		
	1785-1755		
	1020-980		

Table 4.12. Ether Type Compounds

Functional Group	Absorption Range (cm^{-1})	Example (cm^{-1})	Example Compound
ETHERS **Aliphatic**			
O-CH$_3$	2830-2815		
C-O-C	1150-1060	1140	
O-(CH$_2$)$_4$	742-734		Diethyl ether
O-CH$_3$	1455		
Aromatic			
=C-O-C	1275-1200	1247	
C-O-C	1075-1020	1038	Anisol
Cyclic			
C-O-C	1140-1070		
Epoxides			
general	1260-1240	1261	1;2-Epoxybutane
trans compounds	890		
cis compounds	830	826	1;2-Epoxybutane
Tetrahydrofuran derivatives			
	1098-1075	1076	Tetrahydrofuran
	915-913	912	
Trioxans			
	1175	1172	Trioxan
	958	957	
Tetrahydropyran derivatives			
	1120-1080		
	1100-900		
	825-805		
Dioxan derivatives			
	1125	1122	Dioxan
KETALS, ACETALS			
R$_2$-C-(O-C)$_2$	1190-1158		
	1143-1124		
	1098-1063		

Table 4.13. Ketone and Aldehyde Compounds

Functional Group	Absorption Range (cm^{-1})	Example (cm^{-1})	Example Compound
KETONES			
Aliphatic			
	1725-1705	1727	Butanone
	1325-1215	1269	
	1200	1215	
Unsaturated			
C=C	1650-1620	1618	Methyl vinyl ketone
C=O	1685-1665	1684	
Aromatic			
Aryl, alkyl	1700-1680	1694	Acetophenone
Aryl, aryl	1670-1660		
Cyclic			
4- & 5-membered rings	1775-1740	1739	Cyclopentanone
6- & 7-membered rings	1725-1700	1703	Cycloheptanone
Diketones			
α-Diketones	1730-1710	1721	Diacyl ketone
β-Diketones	1640-1540		
γ-Diketones	1725-1705		
Halogen substituted			
α,α-Dihalogen substitution	1765-1745		
α-Dihalogen substitution	1745-1725		
ALDEHYDES			
General			
CH	2900-2700 2 band		
	2720-2700		
	975-780		
Aliphatic			
C=O	1740-1720	1735	Butyraldehyde
CH	1440-1325	1390	
Unsaturated			
C=O	1650-1620	1637	Crotonaldehyde
C=O α,β unsaturated	1690-1650		

Table 4.14. Carboxylic Acid Compounds

Functional Group	Absorption Range (cm^{-1})	Example (cm^{-1})	Example Compound
CARBOXYLIC ACIDS			
General			
OH	3200-2500 br		
CH	1440-1396		
	1320-1210		
OH dimer	950-900 br		
C=O halogen substitution	1740-1720		
C=O aliphatic	1720-1700	1718	Acetic acid
C=O unsaturated	1710-1690	1698	Crotonic acid
C=O aromatic	1700-1680	1695	Benzoic acid

Table 4.14. (Continued)

C=C	1660-1620	1655	Crotonic acid
Carboxylic Ions			
C=O	1610-1560		
C=O	1420-1300		

Table 4.15. Ester Compounds

Functional Group	Absorption Range (cm⁻¹)	Example (cm⁻¹)	Example Compound
ESTERS			
C=O unsaturated, aryl	1800-1770		
C=C unsaturated, aryl	1730-1710	1718	
C-O acrylates, fumarate	1300-1200	1282	Ethyl acetate
C-O	1190-1130		
C=O electronegatively substituted	1770-1745		
C=O α,γ keto	1755-1740		
C=O saturated	1750-1735	1744	Methyl acetate
C=O β keto	1660-1640		
C-O benzoates, phthalates	1310-1250	1277	Methyl benzoate
	1150-1100	1108	
C-O acetates	1250-1230	1246	Propyl acetate
	1060-1000	1047	
C-O phenolic acetates	1205		
C-O formate	1200-1180	1190	Propyl formate
LACTONES			
β-Lactones	1840-1800		
γ-Lactones	1780-1760	1776	Butyrolactone
δ-Lactones	1750-1730		
	1280-1150	1168	

Table 4.16. Anhydride Compounds

Functional Group	Absorption Range (cm⁻¹)	Example (cm⁻¹)	Example Compound
ANHYDRIDES			
Aliphatic			
C=O	1850-1800	1842	Acetic acid anhydride
C=O	1785-1760	1783	
C-O	1170-1050	1134	
Aromatic			
C=O	1880-1840	1866	Phthalic acid anhydride
C=O	1790-1770	1773	
C-O	1300-1200	1267	
Cyclic			
C=O	1870-1820	1818	Glutaric acid anhydride
C=O	1800-1750	1772	

Table 4.17. Amide Compounds

Functional Group	Absorption Range (cm^{-1})	Example (cm^{-1})	Example Compound
AMIDES			
Primary			
NH free	3500		
NH free	3400		
NH bridges	3350	3346	Butyramide
NH bridged	3190	3191	
C=O	1660-1640	1660	
	1430-1400	1430	
Secondary			
NH free *trans*	3460-3400		
NH free *cis*	3440-3420		
NH bridged *trans*	3320-3270	3280	N-Methylacetamide
NH bridged *cis*	3180-3140		
bridged *cis, trans*	3100-3070	3090	
C=O	1680-1630	1652	
NH	1570-1510	1564	
	720 br	725	
Tertiary			
C=O	1670-1630	1670	N,N-Dimethyl formamid

Table 4.18. Amino Acid Compounds

Functional Group	Absorption Range (cm^{-1})	Example (cm^{-1})	Example Compound
AMINO ACIDS			
NH	3130-3030 br		
	2760-2530		
	2140-2080		
C=O	1720-1680		
ionized form	1600-1560		
	1300		
C=O α-amino acids	1754-1720		
C=O β,γ-amino acids	1730-1700		
Amino acid HCl	3030-2500		
NH amino acid HCl's	1660-1590		
NH amino acid HCl's	1550-1490		

Table 4.19. Amine Compounds

Functional Group	Absorption Range (cm^{-1})	Example (cm^{-1})	Example Compound
AMINES			
General			
N-CH$_3$	2820-2730		
N-CH$_3$	1426		
C-N	1410		

Table 4.19. (Continued)

Aliphatic, primary			
NH free	3500-3200	3350	Ethylamine
	(2 bands)	3210	
NH	1650-1590	1630	
	1200-1150		
	1120-1030	1100	
Aliphatic, secondary			
NH free	3500-3200 1 band	3230	Dipropylamine
NH	1650-1550		
C-N	1200-1120	1126	
C-N	1150-1080	1090	
Aliphatic, tertiary			
C-N	1230-1130	1175	Ethyldimethylamine
C-N	1130-1030	1070	
Aromatic, primary			
	3510-3450	3460	Aniline
	3420-3380	3413	
	1630-1600	1621	
Aromatic, secondary			
Free	3450-3430		
Bridged	3400-3300	3400	N-Methylaniline

Table 4.20. Unsaturated Nitrogen Compouds

Functional Group	Absorption Range (cm^{-1})	Example (cm^{-1})	Example Compound
UNSATURATED NITROGEN COMPOUNDS			
Imines			
NH	3400-3300		
C=N	1690-1640		
Oximes			
Liquid	3602-3590		
Solid	3250		
Solids	3115		
Aliphatic	1680-1665		
Aromatic	1650-1620		
	1300		
	900		

Table 4.21. Cyanide and Isocyanide Compounds

Functional Group	Absorption Range (cm^{-1})	Example (cm^{-1})	Example Compound
CYANIDES, ISOCYANIDES			
C≡N unconjugated	2265-2240	2256	Ethyl cyanide
C≡N conjugated or aromatic	2240-2220	2222	Benzyl cyanide
C≡N cyanide,			
thiocyanide complex	2200-2000		
N≡C alkyl isocyanide	2183-2150	2166	Methyl isocyanide
N≡C aryl isocyanide	2140-2080	2100	Phenyl isocyanide

Table 4.22. Cyclic Nitrogen Compounds

Functional Group	Absorption Range (cm^{-1})	Example (cm^{-1})	Example Compound
CYCLIC NITROGEN COMPOUNDS			
Pyridines, quinolines			
CH	3100-3000	3030	Pyridine
C=C, C=N	1615-1590	1590	
	1585-1550		
	1520-1465	1490	
	1440-1410		
	920-690	707	
Pyrimidines			
CH	3060-3010		
C=C, C=N	1580-1520		
Ring	1000-900		

Table 4.23. Unsaturated Nitrogen-Nitrogen Compounds

Functional Group	Absorption Range (cm^{-1})	Example (cm^{-1})	Example Compound
UNSATURATED NITROGEN-NITROGEN COMPOUNDS			
Azo compounds	1630-1575		
N=N azides	2160-2120	2130	Phenylazide
N=N azides	1340-1180	1297	

Table 4.24. Nitro Compounds

Functional Group	Absorption Range (cm^{-1})	Example (cm^{-1})	Example Compound
NITRO COMPOUNDS			
Aliphatic			
	1570-1500	1546	2-Nitrobutane
	1385-1365	1362	
	880	879	
Aromatic			
	1550-1510	1527	Nitrobenzene
	1370-1330	1351	
	849	853	

Table 4.25. Phosphorus Compounds

Functional Group	Absorption Range (cm^{-1})	Example (cm^{-1})	Example Compound
PHOSPHORUS COMPOUNDS			
O-H phosphoric acids	2700-2560 br		
P-H	2440-2350 sp		
P=O	1350-1250		
P=O	1250-1150		
P-O-C	1240-1190		
P-O-R	1190		
P-O-C	1170-1150		

Table 4.25. (Continued)

P-O-C	1050-990		
P-O-P	970-940		
P-F	885		
P=S	840-600		
O-P-H	865-840		
O-P-O	590-520		
O-P-O	460-440		
PHOSPHORUS-CARBON COMPOUNDS			
P-C aromatic	1450-1435		
P-C aliphatic	1320-1280	1298	Trimethylphosphine
P-C	750-650	707	
PO_4^{-3} aryl phosphates	1080-1040		
PO_4^{-3} alkyl phosphates	1180-1150		
PO_4^{-3} alkyl phosphates	1080		

Table 4.26. Deuterated Compounds

Functional Group	Absorption Range (cm^{-1})	Example (cm^{-1})	Example Compound
DEUTERATED COMPOUNDS			
O-D deuterated alcohols	2650-2400		
O-D deuterated carboxylic acids	675		

Table 4.27. Sulfur Compounds

Functional Group	Absorption Range (cm^{-1})	Example (cm^{-1})	Example Compound
SULFUR COMPOUNDS			
C=S	1400-1300	1357	Dithioacetic acid
S=S	1200-1050		
P=S	840-600		
SH mercaptans	2600-2550	2580	Ethyl mercaptan
C-S mercaptans	700-600	665	
C-S-C dialkyl sulfides	750-600	726	Methyl ethyl sulfide
	710-570	676	
	660-630	654	
Aliphatic sulphones	1410-1390	1407	Dimethylsulphone
	1350-1300	1316	
Sulphonic acids	1210-1150		
	1060-1030		
	650		

Table 4.28. Silicon Compounds

Functional Group	Absorption Range (cm^{-1})	Example (cm^{-1})	Example Compound
SILICON COMPOUNDS			
SiH alkylsilanes	2300-2100	2175	Dimethylsilane
$Si(CH_3)_2$	1265-1258	1262	

Table 4.28. (Continued)

Si(CH₃)₃	814-800		
	800		
	1260-1240	1259	Methoxytrimethylsilane
	850-830	844	
	760	763	
Si-C aromatic	1429		
	1130-1090		
Si-C	860-715		
Si-O siloxanes	1100-1000		
Si-O-C open-chain	1090-1020		
Si-O-Si open-chain	1097		
Si-O-Si cyclic	1080-1010		

Table 4.29. Halogen Compounds

Functional Group	Absorption Range (cm⁻¹)	Example (cm⁻¹)	Example Compound
HALOGEN COMPOUNDS			
Iodine Compounds			
	500		
Bromine Compounds			
	700-500		
Chlorine Compounds			
Monochloro	800-600		
	750-700		
Fully chlorinated compounds	780-710		
Fluorine Compounds			
	1400-1000		
	1100-1000		
Fully fluorinated compounds	745-730		

Table 4.30. Inorganic Compounds

Functional Group	Absorption Range (cm⁻¹)	Example (cm⁻¹)	Example Compound
INORGANIC COMPOUNDS			
Sulfates			
	1200-1140	1143	Potassium sulfate
	1130-1080	1117	
	680-610	617	
Nitrates			
	1380-1350	1370	Potassium nitrate
	840-815	825	
Nitrites			
	840-800		
	750		
Water of Crystallization			
	1630-1615		

Table 4.30. (Continued)

Halogen-Oxygen salts			
Chlorates	980-930	978	Potassium chlorate
	930-910	932	
Bromates	810-790	793	Potassium bromate
Iodates	785-730	756	Potassium iodate
Carbonates			
	1450-1410	1410	Calcium carbonate
	880-860	875	

4.4 Nuclear Magnetic Resonance Spectroscopy

4.4.1 Common NMR Solvents

Table 4.31. NMR Solvents

Compound	M.W.	$d^{20/4}$	m.p.[1]	b.p.[1]	δ^2_H(mult)[3]	δ^2_C(mult)[3]
Acetic Acid-d_4	64.078	1.12	17	118	11.53 (1) 2.03 (5)	178 (br) 20.0 (7)
Acetone-d_6	64.117	0.87	-94	57	2.04 (5)	206.0 (13) 29.8 (7)
Acetonitrile-d_3	44.071	0.84	-45	82	1.93 (5)	118.2 (br)
Benzene-d_6	84.152	0.95	5	80	7.15 (br)	128.0 (3)
Chloroform-d	120.384	1.50	-64	62	7.24 (1)	77.0 (3)
Cyclohexane-d_{12}	96.236	0.89	6	81	1.38 (br)	26.4 (5)
Deuterium Oxide	20.028	1.11	3.8	101.4	4.67 (TSP)	
1,2-Dichloroethane-d_4	102.985	1.25	-40	84	3.72 (br)	43.6 (5)
Diethyl-d_{10} Ether	84.185	0.82	-116	35	3.34 (m) 1.07 (m)	65.3 (5) 14.5 (7)
Dimethylformamide-d_7	80.138	1.04	-61	153	8.01 (br) 2.91 (5) 2.74 (5)	162.7 (3) 35.2 (7) 30.2 (7)
Dimethyl-d_6 Sulphoxide	84.170	1.18	18	189	2.49 (5)	39.5 (7)
p-Dioxane-d_8	96.156	1.13	122	101	3.53 (m)	66.5 (5)
Ethyl Alcohol-d_6	52.106	0.91	<-130	79	5.19 (1) 3.55 (br) 1.11 (m)	56.8 (5) 17.2 (7)

Table 4.31. (Continued)

Hexafluoroacetone Deuterate	198.067	1.71	21		5.26 (1)	122.5 (4) 92.9 (7)
HMPT-d$_{18}$	197.314	1.14	7	106	2.53 (2x5)	35.8 (7)
Methyl Alcohol-d$_4$	36.067	0.89	-98	65	4.78 (1)	49.0 (7)
Methylene Chloride-d$_2$	86.945	1.35	-95	40	5.32 (3)	53.8 (5)
Tetrahydrofuran-d$_8$	80.157	0.99	-109	66	3.58 (br) 1.73 (br)	67.4 (5) 25.3 (br)
Trifluoroacetic Acid-d	115.030	1.50	-15	72	11.50 (1)	164.2 (4) 116.6 (4)

1. Melting and boiling points (in °C) are those of the corresponding light compounds. (except for D$_2$O) and are intended only to indicate the useful liquid range of the materials.
2. Chemicals shifts in ppm relative to TMS.
3. The multiplicity br indicates a broad peak without resolvable fine structure, while m indicates one with fine structure.
4. Note that chemical shifts can be dependent on solute, concentration and temperature.

4.4.2 Reference Standards for Proton NMR

Table 4.32. NMR Reference Standards

Compound (Abbrev.)	Formula	m.p. °C	b.p. °C	Hydrogen bands		
				Group[1]	δ	τ
Tetramethylsilane (TMS)[2]	(CH$_3$)$_4$Si		26.5	CH$_3$ (s)	0.00	10.00
Hexamethyl siloxane (HMDS)[3]	[(CH$_3$)$_3$Si]$_2$O	-59	100.4	CH$_3$ (s)	0.04	9.96
Sodium-3-trimethylsilyl-1-propane sulfonate (DSS)[4]	(CH$_3$)$_3$Si-CH$_2$-CH$_2$-CH$_2$-SO$_3$-Na	Solid Salt		CH$_3$ (s) 1-CH$_2$ (m) 2-CH$_2$ (m) 3-CH$_2$ (m)	0.00 0.6 1.8 2.9	10.00 9.4 8.2 7.1
Sodium 3-trimethylsilyl propionate-d$_4$ (TSP)[5]	(CH$_3$)$_3$Si-CD$_2$-CD$_2$-COO-Na	Solid Salt		CH$_3$ (s)	0.00	10.00

1. The functional group which produces the observed band. The multiplicity of the band is indicated in parenthesis; s = singlet; m = complex multiplet.
2. Primary reference standard for room temperature and below.
3. Can be used as reference up to 180° C.
4. Reference for water solutions. The CH$_2$ bands can interfere with weak sample bands.
5. Reference for water solutions.

4.4.3   NMR Proton Chemical Shift Chart

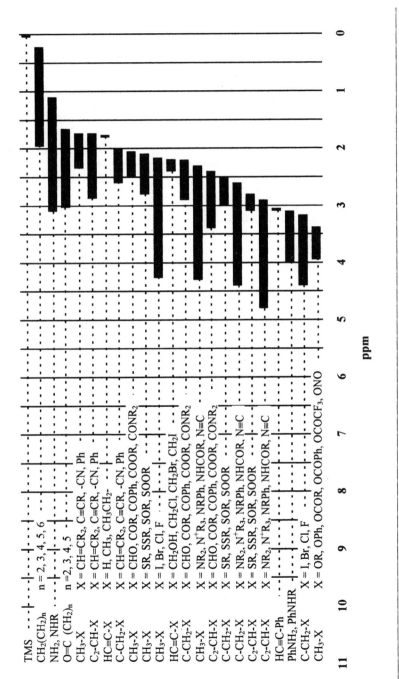

Table 4.33. Proton Chemical Shifts

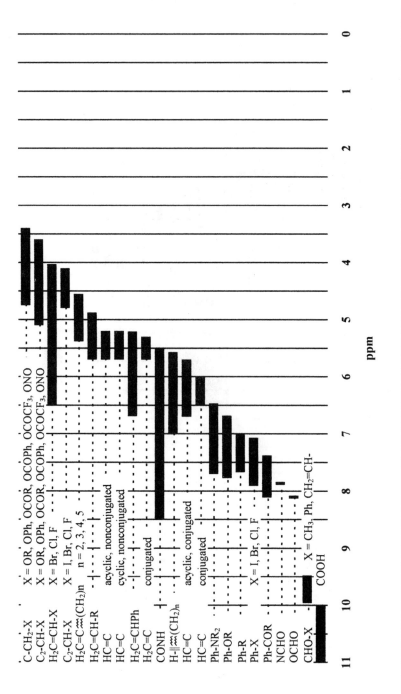

Table 4.33. (Continued)

4.4.4   NMR Chemical Shift Tables

The data in all of the tables are for the compound in dilute carbon tetrachloride or deuteriochloroform relative to internal TMS where such data were available. It is important to remember that solvent effects, especially in the case of aromatics, can cause significant variation in the observed chemical shifts. All chemical shifts listed are in ppm or δ.

Table 4.34. ¹H Chemical Shifts of Paraffinic Compounds with a Single Functional Groups

Group X	CH₃-X	C-CH₂-X	(C)₂-CH-X	CH₃-C-X	C-CH₂-C-X	(C)₂CH-C-X
-H	0.233	0.86	1.33	0.86	1.33	1.56
-CH=CR₂	1.73	2.00	1.73	1.55	1.35	1.00
-C≡CR	1.75	2.15	2.7	1.15	1.50	1.80
-C≡N	1.98	2.35	2.8	1.30	1.6	2.00
-Ph	2.34	2.60	2.87	1.18	1.6	1.8
-CHO	2.17	2.4	2.4	1.13	1.65	---
-COR	2.10	2.4	2.55	1.05	1.5	1.7
-COPh	2.5	2.9	3.4	1.18	1.6	2.0
-CO₂R	2.1	2.2	2.5	1.15	1.7	1.8
-CONR₂	2.05	2.23	2.4	1.1	1.6	1.8
-I	2.16	3.17	4.25	1.8	1.8	2.1
-Br	2.68	3.36	4.2	1.8	1.9	2.0
-Cl	3.05	3.44	4.1	1.5	1.8	2.0
-F	4.26	4.4	4.8	1.4	1.8	2.1
-OR	3.38	3.4	3.6	1.2	---	---
-OPh	3.82	3.95	4.6	1.3	1.5	1.7
-OCOR	3.65	4.1	5.0	1.25	1.6	1.8
-OCOPh	3.82	4.2	5.1	1.5	1.7	1.9
-OCOCF₃	3.95	4.3	---	1.4	1.6	---
-ONO	---	4.75	---	1.4	---	---
-NR₂	2.3	2.6	2.9	1.05	1.45	1.7
-N⁺R₃	~3.2	~3.1	~3.6	1.4	1.7	2.0
-NRPh	~2.7	~3.1	~3.6	1.1	1.5	1.8
-NHCOR	2.8	3.3	3.8	1.1	1.5	1.9
-NO₂	4.30	4.4	4.6	1.6	2.05	2.5
-N≡C	2.85	---	4.8	1.6	---	---
-SR	2.09	2.5	3.0	1.25	1.6	1.9
-SSR	2.30	2.7	---	1.3	1.7	---
-SOR	2.5	3.0	2.8	1.35	1.7	---
-SO₂R	2.8	2.9	3.1	1.35	1.7	---

(From The Chemist's Companion, Gordon & Ford, Copyright © 1972 John Wiley & Sons, Inc. Reprinted with permission of John Wiley & Sons, Inc.)

Table 4.35. [1]H Chemical Shifts of Paraffinic Compounds with Two Functional Groups

Y\X	CN	CF$_3$	Ph	C≡C	C=C	CH$_3$
CH$_3$	2.31	(1.84)	2.63	2.14	1.97	1.34
C=C	3.15	(2.69)	3.30	3.39	2.73	
C≡C	(3.37)	(2.81)	(3.52)	(3.11)		
Ph	3.68	3.50	3.92			
CF$_3$	(3.07)	(2.51)				
CN	4.13					

Y\X	COPh	CONR$_2$	COOR	COR
COR	(3.77)	3.52	3.32	3.62
COOR	(3.62)	(4.95)	4.22	
CONR$_2$	(3.66)	3.30		
COPh	(3.91)			

Y\X	OCOR	OPh	OR	OH
OH	(5.92)	(6.02)	(5.15)	(5.35)
OR	(5.72)	(5.82)	4.49	
OPh	(6.59)	(6.69)		
OCOR	(6.49)			

Y\X	I	Br	Cl
Cl	4.99	5.16	5.28
Br	(4.38)	4.94	
I	3.89		

Y\X	NHCOR	N$_3$	NR$_2$
NR$_2$	(4.07)	(3.77)	3.10
N$_3$	(4.47)	(4.17)	
NHCOR	(4.77)		

Values in parentheses were calculated by the emperical method of Shoolery. All values refer to methylene (CH$_2$) protons in X-CH$_2$-Y.

Table 4.36. [1]H Chemical Shifts of Olefinic Compounds

Compound	X=	a	b	c
H$_2$C=CH$_2$		5.33		
Hb, Ha, Hc, X	R	5.70	4.88	4.96
	F	6.17	4.03	4.37
	Cl	6.30	5.44	5.52
	Br	6.49	5.88	6.03
	C$_6$H$_5$	6.69	5.21	5.71
	CCl$_3$	6.41	5.30	5.78
	CN	5.53	6.05	5.78
	CH$_2$OH	6.00	5.13	5.25
	OCH$_3$	6.43	3.90	4.04

Table 4.36. (Continued)

$H_2C=C=CH_2$		4.55		
$\begin{smallmatrix} Hb \\ \\ Hb \end{smallmatrix} C=C=C \begin{smallmatrix} Ha \\ \\ X \end{smallmatrix}$	Cl Br I	5.76 5.85 5.62	5.05 4.82 4.46	
(1,3-butadiene structure: Hb, Hb, Ha)		5.20	5.11	
(methylenecyclopropane) =CH₂		5.38		
(methylenecyclobutane) CH₂		4.70		
(methylenecyclopentane) =CH₂		4.82		
(methylenecyclohexane) =CH₂		4.55		
CH_3-CHO Ph-CHO $CH_2=CH$-CHO $(CH_3)_2N$-CHO CH_3O-CHO		9.72 9.96 9.48 7.84 8.08		

Table 4.37. ^1H Chemical Shifts of Acetylenic Compounds

Compound	δ	Compound	δ
$H-C{\equiv}C-H$	1.80	$OHC-C{\equiv}C-H$	1.89
$CH_3-C{\equiv}C-H$	1.80	$HOCH_2-C{\equiv}C-H$	2.33
$CH_3CH_2-C{\equiv}C-H$	1.76	$Cl-CH_2-C{\equiv}C-H$	2.40
$Ph-C{\equiv}C-H$	3.05	$Br-CH_2-C{\equiv}C-H$	2.33
$CH_2=CH-C{\equiv}C-H$	2.92	$I-CH_2-C{\equiv}C-H$	2.19
$C_2H_5-C{\equiv}C-C{\equiv}C-H$	1.95	$CH_3O-C{\equiv}C-H$	1.33
$CH_3(C{\equiv}C)_2C{\equiv}C-H$	1.87	$CH_2=CH-O-C{\equiv}C-H$	1.89

Table 4.38. ^{1}H Chemical Shifts of Cycloparaffinic Compounds

Compound	δ	Compound	δ	Compound	δ
Cyclopropane	0.22	Cylcobutane	1.96	Cyclopentane	1.51
Cyclohexane	1.44	Cycloheptane	1.54	Cyclooctane	1.54
Adamantane	1.78				
(bicyclo[4.1.0] structure, H H)	0.02	(decahydronaphthalene, H)	~1.4	(hydrindane, H)	2.37
(bicyclo[2.2.2]octane, Ha, Hb)	a 1.51 b ~2.1	(norbornane, Ha Hb Hc Hd)	a 1.56 b 0.87 c 2.49 d 1.58	(bicyclo[2.2.1] Ha Hb Hc Hd)	a 1.21 b 1.49 c 1.18 d 2.20
(cyclopropanone, =O)	1.65	(cyclobutanone, O, Ha Hb)	a 3.03 b 1.96	(cyclopentanone, O, Ha Hb)	a 2.06 b 2.02
(cyclohexanone, O, Ha)	2.25	(methylenecyclopropane, =CH$_2$, Ha)	0.99	(methylenecyclobutane, =CH$_2$, Ha Hb)	a 2.70 b 1.92
(methylenecyclopentane, =CH$_2$, Ha Hb)	a 2.70 b 1.92	(methylenecyclohexane, =CH$_2$, Ha)	1.5		

Table 4.39. ^{1}H Chemical Shifts Of Cycloolefinic Compounds

Compound	δ	Compound	δ	Compound	δ
(cyclopropene, Ha Hb)	a 0.92 b 7.01	(cyclobutene, Ha Hb)	a 2.57 b 5.97	(cyclopentene, Ha Hb)	a 2.28 b 5.60

Table 4.39. (Continued)

(cyclopentadiene structure) H	6.42	(cyclohexene structure) Ha / Hb	a 1.96 b 5.57	(cyclohexadiene structure) H	2.15
(norbornadiene CH₂ structure) Ha / Hb	a 1.95 b 3.53	(norbornene structure) H	5.95	(benzonorbornene structure) H	6.66
(bicyclic structure) H	6.25	(bicyclic structure) H	6.27	(bicyclic structure) H	6.70

Table 4.40. ¹H Chemical Shifts Of Monosubstituted Benzenes

Substituent	Ortho	Meta	Para
H	7.27	7.27	7.27
CH₃	7.07	7.07	7.07
CH₂CH₃	7.13	7.13	7.13
CH₂OH	7.28	7.28	7.28
CH₂Cl	7.32	7.32	7.32
CHCl₂	7.42	7.42	7.42
CCl₃	7.91	7.40	7.37
CH=CH₂	7.5	7.5	7.5
CHO	7.83	7.49	7.56
COCH₃	7.89	7.41	7.56
CO₂H	8.12	7.43	7.51
CO₂CH₃	7.98	7.38	7.48
COCl	8.11	7.49	7.63
COBr	8.07	7.48	7.64
CONH₂	7.8	7.5	7.5
CN	7.63	7.45	7.55
F	6.99	7.24	7.08
Cl	7.30	7.25	7.18
Br	7.45	7.19	7.23
I	7.67	7.06	7.27
NH₂	6.52	7.02	6.62
NHCH₃	6.47	7.05	6.59
N(CH3)₂	6.61	7.09	6.60
NHCOCH₃	7.7	7.1	7.0

Table 4.40. (Continued)

NH_3^+	7.7	7.5	7.5
NO	7.81	7.55	7.61
NO_2	8.22	7.53	7.65
OH	6.68	7.15	6.82
OCH_3	7.77	7.15	7.37
$OCOCH_3$	6.79	7.18	6.83
SCH_3	7.4	7.2	7.1
SO_2Cl	8.04	7.62	7.72
SO_3CH_3	7.87	7.53	7.60

(From The Chemist's Companion, Gordon & Ford, Copyright © 1972 John Wiley & Sons, Inc. Reprinted with permission of John Wiley & Sons, Inc.)

Table 4.41. ^{1}H Chemical Shifts Of Heteroaromatic Compounds

Compound	δ	Compound	δ
	a 7.38 b 6.30		a 7.19 b 7.04
	8.19		8.58
	a 6.62 b 6.05		a 8.88 b 7.41 c 7.98
	a 7.55 b 6.25 c 7.55		a 7.14 b 7.14 c 7.70
	a 8.50 b 7.06 c 7.46		a 6.57 b 7.26 c 6.15 d 7.13

Table 4.41. (Continued)

Structure	δ	Structure	δ
(pyridazine, Ha, Hb)	a 9.17 b 7.68	(pyrimidine, Ha, Hb, Hc)	a 9.15 b 8.60 c 7.09
(pyrazine)	8.5	(1,3,5-triazine)	9.18
(indole, Ha, Hb)	a 6.54 b 6.34	(benzothiazole)	8.95
(indazole, Ha, Hb, Hc, Hd, He)	a 8.03 b 7.77 c 7.12 d 7.34 e 7.58	(purine, Ha, Hb, Hc)	a 8.63 b 9.16 c 8.95
(indolizine, Ha–Hg)	a 7.14 b 6.64 c 6.28 d 7.25 e 6.50 f 6.35 g 7.76	(quinoline, Ha–Hg)	a 8.81 b 7.26 c 8.00 d 7.68 e 7.43 f 7.61 g 8.05
(benzofuran, Ha–Hf)	a 7.52 b 6.66 c 7.49 d 7.13 e 7.19 f 7.42	(acridine, Ha–He)	a 8.20 b 7.69 c 7.39 d 7.80 e 8.53

Table 4.42. ^{1}H Chemical Shifts Of Heterocyclic Compounds

Compound	δ	Compound	δ	Compound	δ
(oxirane)	2.54	(aziridine)	1.48	(thiirane)	2.27

Table 4.42. (Continued)

Structure (labels)	Shifts	Structure (labels)	Shifts	Structure (labels)	Shifts
oxetane (Hb, Ha)	a 4.73, b 2.72	azetidine (N–H, Hb, Ha)	a 3.54, b 2.23	thietane (Hb, Ha)	a 2.82, b 1.93
tetrahydrofuran (Ha, Hb)	a 3.63, b 1.79	pyrrolidine (N–H, Ha, Hb)	a 2.74, b 1.62	tetrahydrothiophene (Ha, Hb)	a 2.82, b 1.93
tetrahydropyran (Ha, Hb, Hb, Hb)	a 3.56, b 1.58	piperidine (N–H, Ha, Hb, Hb)	a 2.69, b 1.49	thiane (Ha, Hb, Hb)	a 2.57, b 1.60
β-lactone (Hb, Ha)	a 3.48, b 4.22	γ-butyrolactone (Ha, Hc, Hb)	a 2.31, b 2.08, c 4.28	δ-valerolactone (Ha, Hc, Hb, Hb)	a 2.27, b 1.62, c 4.06
1,3-dioxolane (Ha, Hb)	a 4.77, b 3.77	1,3-dioxane (Ha, Hb, Hc)	a 4.82, b 3.80, c 1.68	1,4-dioxane	3.59
1,3,5-trioxane	5.00	1,3-dithiane	3.69	1,3,5-trithiane	4.18
SO₂ ring (Ha, Hb)	a 2.92, b 2.16	morpholine (Ha, Hb, N–H)	a 3.57, b 2.83	1,4-oxathiane (Ha, Hb)	a 3.88, b 2.57

Table 4.43. [1]H Chemical Shifts of Hydrogen Bonded To Oxygen, Nitrogen, Sulfur

Compound	δ	Compound	δ
ROH (monomeric)[a]	0.5	Oximes	7-11
ROH (hydrogen bonded)	0.5-5.0	RNH_2	1.1-1.8
ArOH (monomeric)[a]	4.5	R_2NH	1.2-2.1
ArOH (hydrogen bonded)	4.5-9	$ArNH_2$	3.3-4.0
Enols (intramolecular bonded)	15-19	ArNHR	3.1-3.8
Carboxylic acids	10-13	$RCONH_2$, $ArCONH_2$	5-6.5
Sulfonic acids	11-12	RCONHR, ArCONHR	6-8.2
RSH	1-2	RCONHAr, ArCONHAr	7.8-9.4
ArSH	3-4	R_3N^+H	7.1-7.7

(From The Chemist's Companion, Gordon & Ford, Copyright © 1972 John Wiley & Sons, Inc. Reprinted with permission of John Wiley & Sons, Inc.)

4.5   Mass Spectroscopy

4.5.1   Common Molecular Ions Table

Table 4.44. Common Elemental Compositions of Molecular Ions[1]

m/z	Composition
16	CH_4
17	NH_3
18	H_2O
26	C_2H_2
27	CHN
28	C_2H_4, CO, N_2
30	C_2H_6, CH_2O, NO
31	CH_5N
32	CH_4O, N_2H_4, SiH_4, O_2
34	CH_3F, PH_3, H_2S
36	HCl
40	C_3H_4
41	C_2H_3N
42	C_3H_6, C_2H_2O, CH_2N_2
43	C_2H_5N, N_3H
44	C_2H_4O, C_3H_8, C_2HF, CO_2, N_2O
45	C_2H_7N, CH_3NO
46	C_2H_6O, C_2H_3F, CH_6SI, CH_2O_2, NO_2
48	C_2H_5F, CH_4S, CH_5P
50	C_4H_2, CH_3Cl
52	C_4H_4, CH_2F_2
53	C_3H_3N, HF_2N
54	C_4H_6, F_2O
55	C_3H_5N
56	C_4H_8, C_3H_4O, $C_2H_4N_2$
57	C_3H_7N, C_2H_3NO

Table 4.44. (Continued)

58	C_3H_6O, C_4H_{10}, $C_2H_2O_2$, $C_2H_6N_2$
59	C_3H_9N, C_2H_5NO, CH_5N_3
60	C_3H_8O, C_3H_5F, C_2H_8N2, $C_2H_4O_2$, C_2H_8Si, C_2H_4S, C_2HCl, CH_4N_2O, COS
61	C_2H_7NO, CH_3NO_2, CCIN
62	C_3H_7F, C_2H_7P, $C_2H_6O_2$, C_2H_6S, C_2H_3Cl
64	$C_2H_2F_2$. C_2H_5FO, C_2H_5CL, SO_2
66	C_5H_6, $C_2H_4F_2$, CF_2O, F_2N_2
67	C_4H_5N, CH_3F_2N, ClO_2
68	C_5H_8, C_4H_4O, $C_3H_4N_2$, C_3O_2, CH_2ClF
69	C_4H_7N, C_3H_3NO, $C_2H_3N_3$
70	C_5H_{10}, C_4H_6O, $C_3H_6N_2$. CH_2N_4, CHF_3
71	C_4H_9N, C_3H_5NO, F_3N
72	C_4H_8O, C_5H_{12}, $C_3H_4O_2$, C_4H_5F
73	$C_4H_{11}N$, C_3H_7NO, C_2H_3NS, $C_2H_7N_3$
74	$C_4H_{10}O$, $C_3H_6O_2$, $C_3H_{10}N_2$, C_3H_6S, $C_2H_6N_2O$, $C_3H_{10}Si$, C_3H_3Cl, CH_6N_4. $C_2H_6N_2O$, $C_2H_2O_3$,
75	C_3H_9NO, $C_2H_5NO_2$, C_2H_2ClN
76	$C_3H_8O_2$, C_3H_8S, C_3H_5Cl, C_4H_9F, C_4N_2, C_3H_9P, C_3H_5FO, $C_2H_4O_3$, C_2H_4OS, CH_8Si_2, CH_4N_2S, CS_2
77	CH_3NO_3
78	C_6H_6, C_3H_7Cl, C_4HSN_2, C_2H_6OS, C_2H_3ClO, $C_2H_3FO_2$, CF_2N_2, $CH_6N_2O_2$
79	C_5H_5N
80	C_6H_8, $C_4H_4N_2$, $C_3H_6F_2$, C_2H_5ClO, C_2H_2ClF, CH_4O_2S, HBr
81	C_5H_7N, $C_3H_3N_3$, $C_2H_5F_2N$
82	C_6H_{10}, $C_4H_6N_2$, C_5H_6O, C_2H_4ClF, $C_2H_2N_4$, C_2HF_3, CCIFO
83	C_5H_9N, $C_3H_5N_3$, C_4H_5NO
84	C_6H_{12}, C_5H_8O, $C_4H_8N_2$, $C_4H_4O_2$, $C_2H_4N_4$, C_4H_4S, $C_2H_3F_3$, CH_2Cl_2
85	$C_5H_{11}N$, C_4H_7NO, C_3H_3NS, CH_3N_5
86	$C_5H_{10}O$, $C_4H_6O_2$, C_6H_{14}, $C_4H_{10}N_2$, C_4H_6S, C_4H_3Cl, $C_3H_6N_2O$, $C_2H_2N_2S$, $CHClF_2$, HF_2O, Cl_2O, F_2OS
87	$C_5H_{13}N$, C_4H_9NO, $C_3H_9N_3$, C_3H_5NS, C_3H_2ClN, ClF_2N, F_3NO
88	$C_5H_{12}O$, $C_4H_8O_2$, $C_4H_{12}N_2$, C_4H_8S, $C_3H_8N_2O$, $C_3H_4O_3$, $C_4H_{12}Si$, C_4H_5Cl, CF_4
89	$C_4H_{11}NO$, $C_3H_7NO_2$, C_3H_4ClN
90	$C_4H_{10}O_2$, $C_4H_{10}S$, C_4H_7Cl, $C_3H_6O_3$, $C_3H_{10}OSi$, C_3H_6OS, C_3H_3ClO, $C_2H_6N_2O_2$, $C_2H_6N_2S$, $C_2H_2O_4$, $C_4H_{11}P$
91	$C_2H_5NO_3$, CH_5N_3S, C_3H_6ClN
92	C_7H_8, C_4H_9Cl, C_3H_5ClO, $C_2H_4O_2S$, $C_3H_8O_3$, C_3H_9FSi, $C_5H_4N_2$, C_6H_4O
93	C_6H_7N, C_5H_3NO, $C_4H_3N_3$
94	$C_5H_6N_2$, C_7H_{10}. C_6H_6O, C_3H_7ClO, $C_2H_2S_2$, $C_2H_6O_2S C_3HF_3$, $C_2H_3ClO_2$ C_2Cl_2, CH_3Br
95	C_5H_5NO, C_6H_9N, $C_4H_5N_3$, C_2F_3N
96	C_7H_{12}, C_6H_8O, $C_5H_8N_2$, $C_5H_4O_2$, $C_4H_4N_2O$, $C_2H_2Cl_2$, $C_3H_3F_3$, $C_2H_6F_2Si$
97	C_5H_7NO, $C_6H_{11}N$
98	C_7H_{14}, $C_6H_{10}O$, $C_5H_6O_2$, $C_4H_6O_2$, $C_4H_6N_2O$, $C_3H_6N_4$, $C_5H_{16}N_2$, C_5H_6S $C_4H_2O_3$, $C_2H_4Cl_2$, C_2HClF_2, CCl_2O
99	$C_6H_{13}N$, C_5H_9NO. C_4H_5NS, $C_4H_5NO_2$, CH_3F_2NS

Table 4.44. (Continued)

100	$C_6H_{12}O$, $C_5H_8O_2$, C_7H_{16}, $C_4H_4O_3$, $C_3H_4N_2S$, $C_5H_{12}N_2$, C_6H_9F, $C_5H_{12}Si$, C_4H_4OS, $C_3H_4N_2O_2$, $C_2H_3F_3O$, C_2F_4
101	$C_6H_{15}N$, $C_5H_{11}NO$, $C_4H_7NO_2$, C_4H_7NS, $C_2H_3N_3S$
102	$C_6H_{14}O$, $C_5H_{10}O_2$, $C_5H_{10}S$, $C_4H_{10}N_2O$, $C_4H_6O_3$, $C_2H_2F_4$, $C_3H_6N_2S$, C_8H_6, $CHCl_2F$, CHF_3S, HF_2PS, Cl_2S
103	$C_4H_9NO_2$, C_7H_5N, $C_5H_{13}NO$, $C_4H_{13}N_3$, C_4H_6ClN, $C_2H_2ClN_3$
104	$C_5H_{12}O_2$, $C_5H_{12}S$, C_5H_9Cl, $C_4H_8O_3$, $C_4H_8O_3$, C_4H_8OS, C_8H_8, $C_6H_{13}F$, $C_6H_4N_2$, $C_4H_{12}N_2O$, $C_4H_{12}OSi$, C_4H_5ClO, $C_3H_8N_2S$, $CClF_3$, SiF_4
105	C_7H_7N, $C_3H_7NO_3$, $C_4H_{11}NO_2$, C_4H_8ClN, $CBrN$
106	C_8H_{10}, $C_5H_{11}Cl$, $C_6H_6N_2$, C_7H_6O, $C_4H_{10}O_3$, C_4H_7ClO, $C_4H_{10}OS$, $C_3H_6O_2S$, C_2H_3Br
107	C_7H_9N, C_6H_5NO. $C_2H_5NO_4$
108	$C_6H_8N_2$, C_8H_{12}, C_7H_8O, C_4H_9ClO, C_3H_9ClSi, $C_3H_3ClO_2$, $C_6H_4O_2$, $C_3H_8S_2$, C_2H_5Br, $C_2H_4O_3S$, SF_4
109	C_6H_7NO, $C_7H_{11}N$, $C_2H_4ClNO_2$
110	C_8H_{14}, $C_7H_{10}O$, $C_5H_6N_2O$, $C_3H_4Cl_2$, C_7H_7F, $C_6H_6O_2$, $C_4H_6N_4$, $C_2H_6O_3S$, $C_2H_7O_3P$, $C_3H_7ClO_2$, $C_6H_{10}N_2$
111	$C_7H_{13}N$, C_6H_9NO, $C_5H_5NO_2$, $C_4H_5N_3O$, C_2ClF_2N, C_6H_6FN, $C_5H_9N_3$, C_5H_5NS
112	C_8H_{16}, $C_7H_{12}O$, $C_6H_8O_2$, $C_6H_{12}N_2$, $C_6H_{12}N_2$, C_6H_8S, $C_5H_8N_2O$, $C_5H_4O_3$, $C_4H_4N_2O_2$, $C_3H_6Cl_2$, C_6H_5FO, C_6H_5Cl, C_5H_4OS, C_3F_4, $C_3H_3F_3O$, CH_2BrF
113	$C_7H_{15}N$, $C_6H_{11}NO$, $C_5H_7NO_2$, C_5H_7NS, C_5H_4ClN, $C_3H_3NO_2$, $C_2H_6F_2NP$
114	$C_7H_{14}O$, $C_6H_{10}O_2$, C_8H_{18}, $C_6H_{14}N_2$, $C_4H_6N_2O_2$, $C_5H_6N_2O_2$, C_5H_6OS, $C_6H_4F_2$, $C_5H_2O_3$, C_6H_7Cl, $C_4H_6N_2S$, $C_2H_4Cl_2O$, C_2HCl_2F, C_2HClF_2O, $C_2HF_3O_2$
115	$C_6H_{13}NO$, $C_7H_{17}N$, $C_5H_9NO_2$, C_5H_9NS, C_4H_5NOS, $C_3H_5N_3S$, $C_5H_{13}N_3$
116	$C_6H_{12}O_2$, $C_7H_{16}O$, $C_6H_{12}S$, C_9H_8, $C_6H_{16}N_2$, $C_6H_{16}Si$, C_6H_9Cl, $C_5H_{12}N_2O$, $C_5H_8O_3$, $C_4H_8N_2O_2$, $C_4H_4O_4$, $C_4H_8N_2S$, $C_4H_4S_2$, $C_2H_3Cl_2F$, C_2ClF
117	C_8H_7N, $C_6H_{15}NO$, $C_5H_{11}NO_2$, C_5H_6ClN, $C_4H_7NO_3$, C_3H_4ClN
118	$C_6H_{14}O_2$, $C_6H_{14}S$, $C_5H_{10}O_3$, C_9H_{10}, $C_7H_{16}N_2$, $C_5H_{10}OS$, $C_6H_{15}P$, $C_7H_{15}F$, $C_6H_{11}Cl$, $C_5H_{14}OSi$, $C_4H_{10}N_2O_2$, $C_4H_6O_4$, C_4H_3ClS, $C_4H_{10}N_2S$, $C_4H_{14}Si_2$, C_3H_3Br, $C_2H_2ClF_3$, $CHCl_3$
119	C_7H_5NO, $C_6H_4N_3$, C_8H_9N, $C_4H_9NO_3$, C_4H_9NOS
120	C_9H_{12}, C_8H_8O, $C_4H_8O_2S$, $C_4H_8S_2$, $C_7H_8S_2$, $C_7H_8N_2$, $C_6H_{13}Cl$, $C_5H_{12}O_3$, $C_6H_4N_2O$, $C_5H_4N_4$, $C_4H_8O_4$, C_5H_9OCl, $C_4H_{12}O_2Si$, C_3H_5Br, CCl_2F_2
121	$C_8H_{11}N$, C_7H_7NO, $C_6H_7N_3$, $C_4H_3N_5$, C_7H_3FN
122	$C_8H_{10}O$, $C_7H_{10}N_2$, $C_7H_6O_2$, $C_4H_7ClO_2$, $C_4H_{10}S_2$, C_9H_{14}, C_3H_7Br, C_8H_7F, $C_6H_6N_2O$, $C_4H_{10}O_2S$, $C_4H_4Cl_2$, $C_3H_6O_3S$, $C_2H_6N_2S_2$, C_2H_3BrO
123	C_7H_9NO, $C_6H_5NO_2$, $C_8H_{13}N$, $C_6H_9N_3$, $C_4H_4F_3N$, $C_3H_9NO_2S$
124	$C_3H_8O_3S$, C_8H_9F, $C_7H_{12}N_2$, C_7H_5FO, $C_5H_4N_2O_2$, $C_4H_6F_2O_2$, $C_3H_5ClO_3$, C_2H_5BrO, $C_2H_4S_3$
125	$C_8H_{15}N$, $C_6H_{11}N_3$, C_6H_7NS, $C_7H_{11}NO$, $C_6H_7NO_2$, $C_2H_8NO_3P$, $C_2H_4ClNO_3$
126	C_9H_{18}, $C_8H_{14}O$, $C_4H_8Cl_2$, $C_7H_{10}O_2$, $C_5H_6N_2O_2$, C_7H_7Cl, $C_6H_{10}N_2O$, $C_6H_6O_3$, $C_7H_{10}S$, $C_3H_4Cl_2$, $C_2H_6S_3$, $C_7H_{14}N_2$, C_7H_7FO, C_6H_6OS, $C_4H_6N_4O$, $C_4H_5F_3O$, $C_3H_2N_6$, C_2ClO_2
127	$C_7H_{13}NO$, $C_8H_{17}N$, C_6H_2ClN, $C_5H_9N_3O$, $C_5H_5NO_3$, $C_6H_9NO_2$, $C_4H_5N_3S$, C_2Cl_2FN

Table 4.44. (Continued)

128	$C_8H_{16}O$, C_9H_{20}, $C_7H_{12}O_2$, $C_7H_{12}O_2$, $C_7H_{12}S$, $C_8H_4N_2$, $C_6H_8O_3$, C_6H_6OS, C_6H_5ClO, $C_4H_4N_2OS$, $C_3H_6Cl_2O$, $C_{10}H_8$, $C_2H_6Cl_2Si$, $C_2H_2Cl_2O_2$, CH_2BrCl ,$C_2H_5ClO_2S$, $C_8H_{13}F$, HI
129	$C_8H_{19}N$, $C_7H_{15}NO$, $C_6H_{11}NO_2$, C_9H_7N, $C_7H_3N_3$, $C_6H_{15}N_3$, $C_5H_7NO_3$, C_5H_7NOS, $C_4H_7N_3S$, $C_4H_4ClN_3$, $C_4H_3NO_2S$
130	$C_7H_{14}O_2$, $C_8H_{18}O$, $C_6H_{10}O_3$, $C_6H_6N_2$, $C_{10}H_{10}$, C_9H_6O, $C_7H_{14}S$, $C_6H_{14}N_2O$, C_6H_4ClF, $C_5H_6O_4$, $C_5H_6S_2$, $C_3H_5Cl_2F$, $C_5H_{10}N_2O_2$, $C_3H_6N_4S$, $C_3H_2ClF_3$, $C_3H_2N_2O_2S$, C_2HCl_3, $CHBrF_2$
131	C_9H_9N, $C_7H_{17}NO$, $C_5H_9NO_3$, $C_7H_5N_3$, $C_6H_{17}N_3$, $C_6H_{13}NO_2$, $C_4H_9N_3S$, CF_3NOS
132	$C_7H_{16}O_2$, $C_6H_{12}O_3$, $C_{10}H_{12}$, C_9H_8O, $C_7H_{16}S$, $C_8H_8N_2$, $C_6H_{16}OSi$, $C_5H_8O_4$, $C_6H_{12}OS$, C_6H_9ClO, $C_5H_{12}N_2O_2$, $C_5H_{12}N_2S$, $C_5H_8O_2S$, $C_5H_5ClO_2$, $C_4H_4OS_2$, $C_3H_{12}Si_3$, $C_3H_4N_2S_2$, $C_2H_3Cl_3$, $C_2Cl_2F_2$, $C_2F_4O_2$
133	$C_9H_{11}N$, C_8H_7NO, $C_7H_7N_3$, $C_5H_{11}NO_3$, $C_4H_7NO_2S$, C_3H_4BrN, $C_2H_3ClF_3N$
134	$C_9H_{10}O$, $C_{10}H_{14}$, $C_6H_{14}O_3$, $C_6H_6N_4$, $C_5H_{10}O_2S$, $C_8H_6O_2$, $C_8H_6O_2$, $C_8H_{10}N_2$, C_8H_6S, $C_7H_{15}Cl$, $C_7H_6N_2O$, $C_6H_{14}OS$, $C_6H_{11}ClO$, $C_5H_{11}ClSi$, $C_5H_{10}S_2$, $C_5H_7ClO_2$, $C_3H_3F_5$, $C_3Cl_2N_2$, $C_2H_2Cl_2F_2$
135	$C_9H_{13}N$, C_8H_9NO, C_7H_5NS, $C_5H_5N_5$, $C_7H_5NO_2$, $C_6H_5N_3O$, $C_4H_9NO_2S$, C_3H_6BrN, $C_3F_3N_3$
136	$C_9H_{12}O$, $C_{10}H_{16}$, $C_8H_8O_2$, C_4H_9Br, $C_8H_{12}N_2$, $C_8H_{12}Si$, C_8H_8S, C_8H_5Cl, $C_7H_8N_2O$, $C_7H_4O_3$, $C_5H_{12}S_2$, $C_6H_4N_2O_2$, $C_6H_4N_2O_2$, $C_6H_4N_2S$, $C_5H_{12}O_4$, $C_5H_{12}OS$, $C_5H_9ClO_2$, $C_5H_4N_4O$, C_2HClF_4, CCl_3F
137	$C_8H_{11}NO$, $C_9H_{15}N$, $C_7H_7NO_2$, C_7H_7NS, C_7H_4ClN, $C_6H_7N_3O$, $C_5H_6F_3N$, $C_5H_3N_3S$, $C_3H_7NO_5$
138	$C_{10}H_{18}$, $C_9H_{14}O$, $C_8H_{10}O_2$, $C_8H_{10}S$, $C_7H_6O_3$, $C_6H_6N_2O_2$, C_8H_7Cl, $C_7H_{10}N_2O$, C_7H_6OS, $C_6H_{10}N_4$, $C_5H_6N_4O$, $C_4H_{11}O_3P$, $C_4H_{10}O_3S$, C_3H_7BrO, $C_3H_6S_3$, $C_2H_3BrO_2$, C_2F_6
139	C_7H_9NS, $C_9H_{17}N$, $C_7H_{13}N_3$, $C_7H_9NO_2$, $C_6H_5NO_3$, $C_5H_5N_3O_2$, C_6H_5NOS
140	$C_9H_{16}O$, $C_{10}H_{20}$, C_8H_9Cl, $C_8H_{12}O_2$, $C_8H_{12}S$, $C_7H_8O_3$, $C_5H_{10}Cl_2$, C_7H_8OS, C_8H_9FO, $C_8H_{16}N_2$, $C_8H_6F_2$, C_7H_5ClO, $C_6H_8N_2O_2$, $C_6H_8N_2S$, $C_6H_4O_4$, $C_6H_4S_2$, $C_4H_6Cl_2O$, C_2H_2BrCl
141	$C_8H_{15}NO$, $C_7H_{11}NO_2$, $C_6H_{11}N_3O$, $C_6H_7NO_3$, $C_9H_{19}N$, $C_7H_{15}N_3$, $C_5H_7N_3S$, $C_4H_6F_3NO$, $C_4H_3F_4N$
142	$C_8H_{14}O_2$, $C_9H_{18}O$, $C_{10}H_{22}$, $C_6H_6O_4$, $C_8H_{14}S$, $C_3H_4Cl_2O_2$, C_7H_7ClO, $C_{11}H_{10}$, $C_9H_{15}F$, $C_8H_{18}N_2$, $C_7H_{14}N_2O$, $C_7H_{10}O_3$, $C_6H_{10}N_2O_2$, $C_6H_6O_2S$, $C_5H_6N_2OS$, $C_4H_8Cl_2O$, $C_4H_5ClF_2O$, C_2H_4BrCl, C_2HBrF_2, CH_3I
143	$C_{10}H_9N$, $C_8H_{17}NO$, $C_7H_{13}NO_2$, $C_9H_{21}N$, $C_7H_{10}ClN$, $C_6H_{13}N_3O$, $C_6H_9NO_3$, C_4H_9NOS, $C_5H_9N_3S$, C_2Cl_3N
144	$C_8H_{16}O_2$, $C_9H_{20}O$, $C_9H_8N_2$, $C_7H_{12}O_3$, $C_6H_8O_4$, $C_{10}H_8O$, $C_{11}H_{12}$, $C_8H_{16}S$, $C_8H_{13}Cl$, $C_7H_{16}N_2O$, $C_7H_{12}OS$, $C_6H_9ClN_2$, $C_5H_8N_2O_3$, $C_3H_3Cl_3$
145	$C_{10}H_{11}N$, C_9H_7NO, $C_8H_7N_3$, $C_7H_{15}NO_2$, $C_6H_{11}NO_3$, $C_5H_{11}N_3S$, $C_3H_3N_3O_2S$
146	$C_8H_{18}O_2$, $C_7H_{14}O_3$, $C_8H_{18}S$, $C_6H_{10}O_4$, $C_6H_{10}O_2S$, $C_{11}H_{14}$, $C_{10}H_{10}O$, $C_9H_{10}N_2$, $C_{10}H_7F$, $C_8H_6N_2O$, $C_7H_{18}OSi$, $C_7H_{14}OS$, $C_7H_6N_4$, $C_6H_{14}N_2O_2$, $C_6H_4Cl_2$, $C_5H_6OS_2$, $C_3H_5Cl_3$, $C_3H_3BrN_2$, C_2HCl_3O, $CHBrClF$
147	$C_{10}H_{13}N$, C_9H_9NO, $C_8H_9N_3$, $C_8H_5NO_2$, $C_7H_5N_3O$, $C_6H_{13}NO_3$, $C_5H_9NO_4$, $C_2H_2BrN_3$

Table 4.44. (Continued)

148	$C_{11}H_{16}$, $C_{10}H_{12}O$, $C_9H_8O_2$, $C_7H_8N_4$, $C_9H_{12}N_2$, $C_6H_{12}O_2S$, $C_8H_{17}Cl$, $C_8H_8N_2O$, $C_8H_4O_3$, C_9H_8S, $C_7H_{16}O_3$, $C_7H_{16}OS$, $C_7H_4N_2O_2$, $C_6H_{12}O_4$, $C_6H_4N_4O$, $C_5H_8O_3S$, C_3HF_5O, $C_2H_3Cl_3O$, C_2Cl_3F, $CBrF_3$
149	$C_{10}H_{15}N$, $C_9H_{11}NO$, $C_7H_7N_3O$, $C_8H_{11}N_3$, $C_8H_7NO_2$
150	$C_{10}H_{14}O$, $C_9H_{10}O_2$, $C_6H_{14}S_2$, $C_8H_{10}N_2O$, $C_6H_{11}ClO_2$, $C_5H_{11}Br$, $C_{11}H_{18}$, $C_8H_{14}N_2$, $C_9H_{10}S$, $C_8H_6O_3$, $C_7H_6N_2S$, $C_6H_{14}O_4$, $C_6H_{14}O_2S$, $C_6H_6N_4O$, $C_6H_2F_4$, C_3F_6, $C_2H_2Cl_3F$
151[2]	(see m/z 137) $C_7H_5NO_3$
152	(see m/z 138) C_8H_5ClO, $C_6H_{10}Cl_2$, CCl_4
153	(see m/z 139) $C_9H_{15}NO$, C_7H_5ClNO
154	(see m/z 140; $C_{10}H_8O$ has highest occurrence of data base) C_8H_7ClO, C_2ClO, C_2ClF_5, $C_8H_{10}OS$
155	(see m/z 141) $C_8H_{13}NS$
156	(see m/z 142) $C_7H_5ClO_2$, C_6H_5Br, $C_6H_4O_5$
157	(see m/z 143) C_5H_4BrN
158	(see m/z 144) CF_6S
159	(see m/z 145) $C_6H_9NS_2$, $C_5H_6ClN_3O$
160	(see m/z 146) $C_7H_{20}Si_2$, C_6H_9Br, $C_5H_{16}Si_3$
161	(see m/z 147) $C_6H_5Cl_2N$
162	(see m/z 148) $C_{10}H_7Cl$, $C_8H_6N_2S$, $C_6H_{11}Br$, C_4F_6, $C_6H_4Cl_2O$, $CHBrCl_2$
163	(see m/z 149) $C_7H_9N_5$, C_3H_2BrNS, CCl_3NO_2
164	(see m/z 150) C_9H_8OS, $C_8H_8N_2O_2$, $C_6H_5ClN_2$, $C_7H_{16}O_4$, $C_6H_4N_4O_2$, C_5H_9BrO, $C_5H_8O_2S_2$, C_2Cl_4, $CBrClF_2$
165	(see m/z 137, 151) $C_6H_7N_5O$
166	(see m/z 138, 152) $C_{13}H_{10}$, $C_8H_6O_4$, $C_8H_6O_2S$, $C_8H_6S_2$, $C_6H_6N_4O_2$, $C_6H_6N_4S$, $C_3H_6N_2O_6$, C_3ClF_5, C_3F_6O
167	(see m/z 139, 153) $C_7H_{13}N_5$, $C_7H_5NO_4$, $C_7H_5NS_2$
168	(see m/z 140, 154) $C_{13}H_{12}$, $C_9H_{16}N_2O$, $C_8H_8O_4$, $C_8H_8O_2S$, $C_8H_8S_2$, $C_6H_4N_2O_4$, C_6HF_5, C_3H_5I, $C_2HCl_3F_2$, Cl_4Si
169	(see m/z 141, 155) C_8H_8ClNO, $C_7H_7NO_2S$
170	(see m/z 142, 156) $C_9H_{16}NO_2$, $C_8H_{10}S_2$, $C_3H_6S_4$, $C_2Cl_2F_4$, C_2F_6S
171	(see m/z 143, 157) $C_7H_9NO_2S$, $C_6H_9N_3O_3$, $C_5H_2ClN_3S$
172	(see m/z 144, 158) $C_{12}H_9F$, $C_8H_6Cl_2$, $C_8H_9ClO_2$, $C_7H_8O_3S$, $C_6H_8N_2O_2S$, C_6H_5BrO, $C_6H_4O_6$, CH_2Br_2
	(see m/z 145, 159) C_6H_4ClNOS, $C_5H_8ClN_5$
173	(see m/z 146, 160) $C_{11}H_{10}S$, $C_{10}H_6O_3$, $C_{10}H_6OS$, $C_9H_6N_2O_2$, C_6H_4BrF,
174	C_5F_6
175	(see m/z 147, 161) $C_8H_5N_3O_2$, $C_5H_9N_3S_2$, $C_7H_{10}ClNO_2$

1. Compositions are listed by m/z value, ranked in decreasing order of occurrence probability for compounds in the *Registry of Mass Spectral Data* (Stenhagen *et al.* 1974). Only the more probable combinations of the elements H, C, N, O, F, Si, P, S, Cl, Br and I are included. Note that these are odd-electron ion compositions; many common even-electron fragment ions have compositions differing by ±1 hydrogen atom, and can therefore be found ±1 mass unit from those listed. The above table can also be used to suggest possible elemental compositions of fragment ions

2. For masses above 150 the only compositions included are those for which a corresponding composition differing by a CH_2 less unit.

Chapter 5

Units and Measurements

5.1 Fundamental Physical Constants
5.2 Units
5.3 Prefixes
5.4 Conversion Factors

5.1   Fundamental Physical Constants

Table 5.1. Physical Constants

Symbol	Name	Value	Units
amu	Atomic Mass Unit	$1.6605655 \times 10^{-24}$	g
N_o	Avogadro's Number	6.022045×10^{23}	units/mole
K	Boltzman's Constant	1.380663×10^{-16}	erg/^{o}K
e	Electron Charge	$1.6021892 \times 10^{-19}$	coulomb
m_e	Electron Rest Mass	9.109534×10^{-28}	g
		5.485803×10^{-4}	amu
eV	Electron Volt	1.60×10^{-19}	joule
F	Faraday's Constant	9.648456×10^4	coulombs/equiv
		2.8925342×10	cal/volt
R	Gas Constant	8.2056×10^{-2}	1atm/^{o}K-mole
		8.3144	joules/^{o}K-mole
		8.3144×10^7	erg/^{o}K-mole
		1.9872	cal/^{o}K-mole
m_n	Neutron Rest Mass	$1.6749543 \times 10^{-24}$	g
		1.0086650	amu
h	Planck's Constant	6.626176×10^{-27}	erg/sec
m_D	Proton Rest Mass	$1.6726485 \times 10^{-24}$	g
		1.0072674	amu
R*	Rydberg Constant	1.0973718×10^5	cm
c	Speed of Light (in vacuum)	2.9979246×10^{10}	cm/sec
atm	Standard Pressure	101.3	kPa
		760	mmHg

5.2   Units

5.2.1   Base SI Units

Table 5.2. Base SI Units

Symbol	Physical Quantity	Name of Base Unit
m	Length	Meter
kg	Mass	Kilogram
s	Time	Second
A	Electric Current	Ampere
K	Thermodynamic Temperature	Kelvin
mol	Amount of Substance	Mole
cd	Luminous Intensity	Candela

5.2.2   Derived SI Units

Table 5.3. Common Derived SI Units

Symbol	Physical Quantity	Name of Unit	Definition of Unit
Å	length	Angstrom	10^{-10}m
μ	length	Micron	10^{-6}m
dyn	force	Dyne	10^{-5}N
bar	pressure	Bar	10^{-5}N $\cdot$ m^{-2}
erg	energy	Erg	10^{-7}J

Table 5.4. Derived SI Units with Special Names

Symbol	Physical Quantity	Name of Unit	Definition of Unit
Hz	frequency	Hertz	1/s
J	energy	Joule	N $\cdot$ m
N	force	Newton	kg $\cdot$ m/s^2
W	power	Watt	J/s
Pa	pressure	Pascal.	N/m^2
C	electric charge	Coulomb	A s
V	difference	Volt	W/A
Ohm	electrical resistance	Ohm	V/A

5.2.3   Non SI Units

Table 5.5. Non SI Units

Symbol	Physical Quantity	Name of Unit	Definition of Unit
in	length	inch	$2.54 \cdot 10^{-2}$ m
lb	mass	pound	0.45359237 kg
kgf	force	kilogram-force	9.80665 N
atm	pressure	atmosphere	101.325 N $\cdot$ m^{-2}
torr	pressure	torr	$(101.325/760)$N $\cdot$ m^{-2}
BTU	energy	British Thermal Unit	1055.056 J
kW	energy	kilowatt-hour	3.6 x 106 J
cal$_{th}$	energy	thermochemical calorie	4.184 J

Table 5.5. (Continued)

eV	energy	electron Volt	$1.60219 \cdot 10^{-19}$ J
amu	mass	atomic mass unit	$1.6605655 \cdot 10^{-27}$ kg
D	electric dipole moment	Debye	$3.3356 \cdot 10^{-30}$ A · m · s
F	charge per molecule	Faraday	$9.648456 \cdot 10^4$ C mol^{-1}

5.3   Prefixes

5.3.1   SI Prefixes

Table 5.6. SI Prefixes

Factor	Prefix	Symbol
10^{15}	penta	P
10^{12}	tera	T
10^9	giga	G
10^6	mega	M
10^3	kilo	k
10^2	hecto	h
10^1	deka	da
10^{-1}	deci	d
10^{-2}	centi	c
10^{-3}	milli	m
10^{-6}	micro	μ
10^{-9}	nano	n
10^{-12}	pico	p
10^{-15}	femto	f

5.3.2   Greek Prefixes

Table 5.7. Greek Prefixes

Value	Prefix
1	mono
2	di
3	tri
4	tetra
5	penta
6	hexa
7	hepta
8	octa
9	ennea
10	deca

5.4   Conversion Factors

5.4.1   Linear Conversion

1 inch = 2.5400 centimeters
1 foot = 0.3048 meter
1 yard = 0.9144 meter
1 mile = 1.6093 kilometers

1 centimeter = 0.3937 inch
1 meter = 3.281 feet
1 meter = 1.0936 yards
1 kilometer = 0.62137 miles

5.4.2   Area Conversion

1 sq. inch = 6.4516 sq. centimeters
1 sq. foot = 0.0929 sq. meter
1 sq. yard = 0.8361 sq. meter
1 sq. mile = 2.59 sq kilometers

1 sq. centimeter = 0.155 sq. inch
1 sq. meter = 10.764 sq. feet
1 sq. meter = 1.196 sq. yards
1 sq. kilometers = 0.3861 sq. mile

5.4.3   Cubic Conversion

1 cu. inch = 16.3872 cu. centimeters
1 cu. foot = 28.317 cu. centimeters
1 cu. yard = 0.7645 cu. meter

1 cu. centimeter = 0.0610 cu. inch
1 cu. decimeter = 0.0353 cu. foot
1 cu. meter = 1.3079 cu. yards

5.4.4   Capacity Conversion

1 fluid ounce = 29.5730 milliliters
1 liquid pint = 0.4732 liter
1 liquid quart = 0.9463 liter
1 gallon = 3.7853 liters
1 dry quart = 1.1012 liters

1 milliliter = 0.0338 fluid ounce
1 liter = 2.1134 fluid pints
1 liter = 1.0567 liquid quarts
1 liter = 0.2642 gallon
1 liter = 0.9081 dry quart

5.4.5   Weight Conversion

1 ounce = 28.350 grams
1 pound = 0.4536 kilograms

1 gram = 0.0353 ounce
1 kilogram = 2.2046 pounds

5.4.6   Temperature Conversion

Temperature given in	To Convert to		
	°C	K	°F
°C	°C	°C + 273.15	1.8 °C + 32
K	K - 273.15	K	1.8K - 459.4
°F	0.556 °F - 17.8	0.556 °F + 255.3	°F

Chapter 6

Mathematical Concepts
6.1 Algebraic Formulas
6.2 Plane Figure Formulas
6.3 Solid Figure Formulas

6.1   Algebraic Formulas

6.1.1   Laws of Exponents

$x^m \cdot x^n = x^m{+}^n$ $x^m \div x^n = x^{m-n}$

$(x^m)^n = x^{m \cdot n}$ $(x \cdot y)^m = x^m \cdot y^m$

$(x/y)^m = x^m/y^m$ $x^0 = 1$

$x^{1/n} = \sqrt[n]{x}$ $x^{-m} = 1/x^m$

6.1.2   Laws of Logarithms

$\log (A \cdot B) = \log A + \log B$ $\log \dfrac{A}{B} = \log A - \log B$

$\log A^b = b \log A$ $\log \sqrt[b]{A} = \dfrac{\log A}{b} = \dfrac{1}{b} \log A$

6.1.3   Quadratic Equation

$ax^2 + bx + c = 0$ (Where a $\neq$ 0 and a, b, and c are real numbers)

173

If the roots of $ax^2 + bx + c = 0$ are represented by r_1 and r_2, then:

1. $r_1 = \dfrac{-b + \sqrt{b^2 - 4ac}}{2a}$ and $r_2 = \dfrac{-b - \sqrt{b^2 - 4ac}}{2a}$

2. $r_1 + r_2 = -b/a$

3. $r_1 r_2 = c/a$

4. $x^2 - (r_1 + r_2)x + r_1 r_2 = 0$

   Using the discriminant to determine the nature of the roots of $ax^2 + bx + c = 0$:

5. If $b^2 - 4ac$ is zero or positive, the roots are real.

6. If $b^2 - 4ac$ is negative, the roots are imaginary.

7. If $b^2 - 4ac$ is zero, the roots are equal.

8. If $b^2 - 4ac$ is not zero, the roots are unequal.

9. If $b^2 - 4ac$ is a perfect square, the roots are rational numbers.

10. If $b^2 - 4ac$ is positive and not a perfect square, the roots are irrational numbers.

6.1.4   Graphs (a, b, c, m, and r are real numbers)

1. The slope of a line that passes through two points $P_1(x_1,y_1)$ and $P_2(x_2,y_2)$, $x1 \neq x2$:

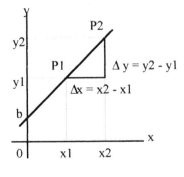

$$m = \text{slope of } P_1P_2 = \frac{y_2 - y_1}{x_2 - x_1} = \frac{\Delta y}{\Delta x}$$

2. Equation of a straight line: $y = mx + b$ where m = slope of the line and b = y intercept.

3. Equation of a parabola: $y = ax^2 + bx + c$ or $x = ay^2 + by = c$, $a \neq 0$.

4. Equation of a circle: $(x-h)^2 + (y-k)^2 = r^2$ where the center is (h, k) and the radius is r, $r > 0$.

5. Equation of a circle: $x^2 + y^2 = r^2$ where the center is at the origin and the radius is r, $r > 0$.

6. Equation of an ellipse: $ax^2 + by^2 = c$ where the center is at the origin; a, b, and c are positive; $c \neq 0$.

7. Equation of a hyperbola: $ax^2 - by^2 = c$, $ay^2 - bx^2 = c$ where the center is at the origin; a and b are positive. Also $xy = k$, k is a constant.

8. At the turning point of the parabola $y = ax^2 + bx + c$, $x = -b/2a$

9. The graph of the parabola $y = ax^2 + bx + c$ opens upward and has a minimum turning point when a is positive, $a > 0$.

10. The graph of the parabola $y = ax^2 + bx + c$ opens downward and has a maximum turning point when a is negative, $a < 0$.

6.2   Plane Figure Formulas

6.2.1   Rectangle

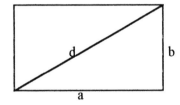

Area $A = ab$

Perimeter $P = 2\,(a + b)$

Diagonal $d = \sqrt{a^2 + b^2}$

If $a = b$ then it is a square

6.2.2   Parallelogram

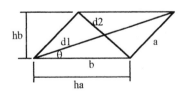

Area $A = ah_a = bh_b = ab \sin \theta$

Perimeter $P = 2(a + b)$

Diagonal:

$d1 = \sqrt{a^2 + b^2 - 2ab\cos\theta}$

$d2 = \sqrt{a^2 + b^2 + 2ab\cos\theta}$

6.2.3   Trapezoid

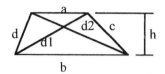

Area $A = \frac{(a+b)\,h}{2}$

Perimeter $P = a + b + c + d$

Diagonal $= d_1 = \sqrt{ab + \frac{ac^2 - bd^2}{a - b}}$

$d_2 = \sqrt{ab + \frac{ad^2 - bc^2}{a - b}}$

Height $= h = \frac{2}{a - b} \cdot \sqrt{s\,(s - a + b)\,(s - c)\,(s - d)}$

where $s = \frac{1}{2}\,(a - b + c + d)$

6.2.4   Equilateral Triangle

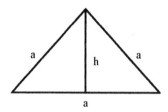

a = all sides equal

Area $A = a^2\,\frac{\sqrt{3}}{4} = 0.433\,a^2$

Perimeter $P = 3\,a$

$h = \frac{\sqrt{3}}{2}\,a = 0.866\,a$

6.2.5   Circle

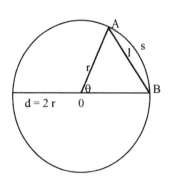

Area of circle $A = \pi\,r^2 = 3.14159\,r^2$

$= \pi\,\frac{d^2}{4} = 0.78539\,d$

Area of sector $AsB0A = \frac{\pi\,\theta\,r^2}{360}$

Area of segment $AsBlA = \frac{\pi\,r^2\,\theta}{360} - \frac{r^2\,\sin\,\theta}{2}$

Circumference $= 2\,\pi\,r = 6.28318\,r$

$= \pi\,d = 3.14159\,d$

Length of chord $AlB = 2\,r\,\sin\,\frac{\theta}{2}$ Length of arc $AsB = \frac{\pi\,r\,\theta}{180}$

6.2.6 Ellipse

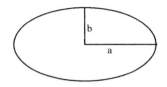

Area $A = \pi\, ab = 3.14159\, a\, b$

Circumference $\approx 2\,\pi\sqrt{\dfrac{a^2+b^2}{2}}$

6.2.7 Parabola

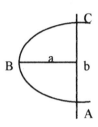

Area ABCA $=2/3\ a\ b$

Length of arc ABC $= b\ \{\tfrac{1}{2}\ (1 + 16\ (a/b)^2)^{\tfrac{1}{2}} +$

$\dfrac{1}{8\left(\tfrac{a}{b}\right)}\ \ln\ [4\ n + (1 + 16\ (a/b)^2)^{\tfrac{1}{2}}]\}$

6.3 Solid Figure Formulas

6.3.1 Parallelpiped

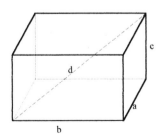

Surface area $A = 2\ (ab + bc + ca)$

Volume $V = abc$

Diagonal $d = \sqrt{a^2 + b^2 + c^2}$

In the case of a cube, $a = b = c$ then;

Surface $= 6\ a^2$

Volume $V = a^3$

Diagonal $d = a\sqrt{3}$

6.3.2   Right Cylinder

Surface area of convex surface $A_c = 2 \pi r h$

$$= 6.283 \, r \, h$$

Total surface area $A = 2 \pi r (r + h)$

Volume $V = \pi r^2 h$

6.3.3   Right Cone

Surface area of convex surface $A_c = \pi r l$

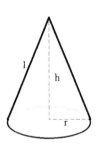

$(l = \text{slant height} = \sqrt{r^2 - h^2}\,)$

Total surface area $A = \pi r (r + l)$

Volume $V = 1/3 \, \pi r^2 h$

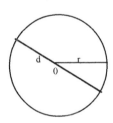

6.3.4   Sphere

Surface area $A = 4 \pi r^2 = \pi d^2$

Volume $V = 4/3 \, \pi r^3 = 1/6 \, \pi d^3$

Index